Spon's Construction Cost and Price Indices Handbook

OTHER TITLES FROM E & F N SPON

Spon's Budget Estimating Handbook
Edited by Spain and Partners

Spon's Guide to Housing, Construction and Property Market Statistics
M C Fleming

Standard Method of Specifying for Minor Works
3rd Edition
L Gardiner

Spon's Construction Cost and Price Indices Handbook
B A Tysoe and M C Fleming

National Taxation for Property Management and Valuation
A MacLeary

Spon's Construction Output Manual
Edited by T Johnson,
Davis Langdon & Everest

Published Annually

SPON'S CONTRACTORS' HANDBOOKS
Spain & Partners

Electrical Installation

Floor, Wall and Ceiling Finishings

Minor Works, Alterations, Repairs and Maintenance

Painting and Decorating

Plumbing and Domestic Heating

Roofing

SPON'S PRICE BOOKS
Davis Langdon & Everest

Spons' Architects' and Builders' Price Book

Spon's Civil Engineering and Highway Works Price Book

Spon's Mechanical and Electrical Services Price Book

Spon's Landscape and External Works Price Book

For more information about these and other titles published us, please contact:
The Promotion Dept., E & F N Spon, 2-6 Boundary Row, London SE1 8HN

Spon's Construction Cost and Price Indices Handbook

Edited by
Michael C. Fleming
Professor of Economics
Loughborough University

and

Brian A. Tysoe
Chartered Quantity Surveyor
Estates Directorate
Department of Health

First edition

E & FN SPON
An Imprint of Chapman & Hall

London · New York · Tokyo · Melbourne · Madras

UK	Chapman & Hall, 2–6 Boundary Row, London SE1 8HN
USA	Van Nostrand Reinhold, 115 5th Avenue, New York NY10003
Japan	Chapman & Hall Japan, Thomson Publishing Japan, Hirakawacho Nemoto Building, 7F, 1-7-11 Hirakawa-cho, Chiyoda-ku, Tokyo 102
Australia	Chapman & Hall Australia, Thomas Nelson Australia, 102 Dodds Street, South Melbourne, Victoria 3205
India	Chapman & Hall India, R. Seshadri, 32 Second Main Road, CIT East, Madras 600 035

First edition 1991

© 1991 Michael C. Fleming and Brian A. Tysoe

Printed in Great Britain by
T.J. Press (Padstow) Ltd, Padstow, Cornwall.

ISBN 0-419-15330-6   0-442-31418-3 (USA)

Apart from any fair dealing for the purposes of research or private study, or criticism or review, as permitted under the UK Copyright Designs and Patents Act, 1988, this publication may be not be reproduced, stored, or transmitted, in any form or by any means, without the prior permission in writing of the publishers, or in the case of reprographic reproduction only in accordance with the terms of the licenes issued by the Copyright Licensing Agency in the UK, or in accordance with the terms of licences issued by the appropriate Reproduction Rights Organization outside the UK. Enquiries concerning reproduction outside the terms stated here should be sent to the publishers at the UK address printed on this page.

The publisher makes no representation, express or implied, with regard to the accuracy of the information contained in this book and cannot accept any legal responsibility or liability for any errors or omissions that may be made.

British Library Cataloguing in Publication Data

Fleming, M.C.
  Spon's construction cost and price indices handbook
  I. Title   II. Tysoe, B.A.
  692.029

  ISBN 0-419-15330-6

Library of Congress Cataloging-in-Publication Data

Available

Contents

Preface	vii
Acknowledgements	ix
Abbreviations and Acronyms	x
Introduction	1

PART A CONSTRUCTION INDICES: USES AND METHODOLOGY

1 Uses of Construction Indices	13
2 Problems and Methods of Measurement	31

PART B CURRENTLY COMPILED CONSTRUCTION INDICES

3 Introduction to Current Construction Indices	39
4 Output Price Indices	41
DOE construction output price indices 1970–	41
5 Tender Price Indices	57
5.1 DOE public sector bulding tender price index, 1975–	57
5.2 PSA QSSD index of building tender prices, 1968–	72
5.3 BCIS tender price indices, 1974–	81
5.4 DOE road construction tender price indices, 1970–	102
5.5 DOE price index of public sector housebuilding, (PIPSH), 1964–	114
5.6 Scottish Office housing tender price index, 1970–	124
5.7 D L & E tender price index, 1966–	128
6 Cost Indices	133
6.1 BCIS building cost indices, 1971–	133
6.2 Spon's cost indices	156
(a) Building costs index, 1966–	160
(b) Mechanical services cost index, 1966–	162
(c) Electrical services cost index, 1966–	164
(d) Constructed civil engineering cost index, 1970–	166
(e) Constructed landscaping cost index, 1976–	168

	6.3	'Building' housing cost index, 1973–	171
	6.4	Scottish Office building cost index, 1970–	178
	6.5	ABI/BCIS house rebuilding cost index, 1978–	182
	6.6	Association of Cost Engineers – erected plant cost indices, 1958–	187
	6.7	BMI maintenance cost indices, 1970–	198
	6.8	PSA APSAB cost indices, 1970–	248
7		Comparison and Review of Current Indices	261

PART C HISTORICAL CONSTRUCTION INDICES

8		Introduction to Historical Construction Indices	287
9		Historical Cost and Price Indices	291
	9.1	Jones's selling price of building index, 1845–1922	291
	9.2	Saville's selling price of building index, 1923–1939	296
	9.3	Maiwald's indices of costs for building and other construction, 1945–1938	300
	9.4	Redfern's indices of costs for building and works and for housing, 1839/1850–1953	305
	9.5	CSO CCA index of costs for new building works, 1888–1956	310
	9.6	Board of Inland Revenue (BIR) index of building costs, 1896–1956	314
	9.7	Venning Hope cost of building index, 1914–1975	318
	9.8	Banister Fletcher's index of the comparative cost of building, 1920–1939	323
	9.9	BRS measured work index, London area, 1939–1969 Q2	328
	9.10	DOE cost of new construction (CNC) index, 1949 Q1–1980 Q1	332
10		Comparison and Review of Historical Indices	339

Bibliography	349
Appendix A General Indices of Prices	351
A1 Cost of living index, 1914–1948	352
A2 Index of retail prices, 1948–1989	353
A3 Index of total home costs, 1948–1989	354
A4 Index of capital goods prices, 1948–1989	355
Appendix B Names, Addresses and Telephone Numbers	357
Glossary	359
Index	365

Preface

This handbook brings together all of the major series of index numbers which measure changes over time in construction costs and prices in the United Kingdom. In total over ninety series are included, covering over eighty series which are currently compiled and ten historical series (extending back to the first half of the nineteenth century). Coverage does not encompass the wide range of information available about house prices (that information is covered in *Spon's House Price Data Book*) but indices of housebuilding *costs* are included.

The book is designed to be a convenient source of reference. In writing it we have had the needs of quantity surveyors, both practitioners and students, primarily in mind. But the focus is not meant to be narrow. It is in fact aimed at all those who have a need for information about movements in the costs of building and other construction work, including not only quantity surveyors but also valuers, architects, civil and structural engineers and others who provide professional advice on construction costs as well as developers and contractors themselves. It will also be useful to those concerned with the valuation of construction assets for insurance or company valuation purposes, to statisticians concerned with the compilation of statistics of the value of construction output and capital stock at constant prices and to economists concerned with studies of investment, economic growth and inflation.

Index numbers have their limitations and none more so than construction indices. But their proliferation in the single sector of the economy covered by construction may be taken as an indication of the extent of the needs which they have been developed to serve. However, the existence of such a large number of indices, many apparently measuring the same thing, creates problems of choice and many pitfalls for the unwary user. There is, therefore, a great need for guidance in this area and this has motivated the preparation of this handbook. In summary, our intention is twofold. One aim is to provide a convenient and comprehensive source for this vast and confusing range of information. The second aim is to provide a guide through the maze to help users choose the correct index for particular applications and to provide the background information which is essential for an understanding of the scope and limitations of the series available.

The book covers the same subject area as *Construction Cost and Price Indices: Description and Use* by Tysoe, published by Spon in 1981, but in other respects it is different from that volume. Firstly, it has been completely rewritten to take account of the large number of changes that have occurred since 1981. Secondly, it has been expressly designed to make it more useful as a convenient source of reference, hence the use of the word 'handbook' in the title. Thirdly, it has been expanded to include historical series, as indicated above, because the long length of life of most construction assets means that there is a special need for construction indices covering correspondingly long periods of time. Fourthly, the available series are classified systematically by type and each is described following a standard format. Fifthly, the actual index numbers themselves are presented, in both tabular and graphical form, as a continuous series (any breaks in continuity caused by periodic revision of base dates being eliminated by rescaling the whole series to a common base). Lastly, the whole book has been placed within a framework in which the theoretical and practical problems of measuring construction cost and prices movements over time are discussed and the practical application of the indices which are available is explained in detail.

Although designed primarily as a source of reference, summary comparisons are made of the evidence provided by the available indices about construction cost and price movements over time and comparisons are also made with the movement of costs and prices in the economy as a whole.

Every effort has been made to ensure the accuracy of the information given in this book, but neither the authors nor the publishers in any way accept liability for loss of any kind resulting from the use made of this information by any person.

M.C. Fleming, Loughborough
B.A. Tysoe, Harrow

By the same authors

By Michael C. Fleming

Housing in Northern Ireland (1974, Heinemann Educational Books)

Construction and the Related Professions (1980, Pergamon Press)

Statistics Collected by the Ministry of Works 1941-56.
Two volumes (1980, Department of the Environment)

Spon's Guide to Housing, Construction and Property Market Statistics
(1986, E. & F.N. Spon)

Spon's House Price Data Book (with J.G. Nellis) (1987, E. & F.N. Spon)

Essence of Statistics for Business (with J.G. Nellis)
(forthcoming 1991, Philip Allan)

By Brian A. Tysoe

Construction Cost and Price Indices: Description and Use
(1981, E. & F.N. Spon)

Acknowledgements

A large number of organisations are responsible for the calculation of the indices of costs and prices included in this volume. We would like to acknowledge here the permission kindly granted to us by these bodies to reproduce the series for which they are responsible, in particular, the Association of Cost Engineers, the Building Cost Information Service, Building Maintenance Information Limited, Davis, Langdon and Everest and Venning Hope Limited. The indices produced by government and other official bodies are Crown copyright and are reproduced here by permission of the Controller of Her Majesty's Stationery Office. We are most grateful to all of these bodies for their cooperation and for the help they have given us during the preparation of this volume.

We owe a special word of thanks to Stephen Fleming for his work in collating and rebasing some of the statistical series at an early stage of the project. We should also like to express our thanks and admiration for the work carried out by Mrs Su Spencer who typed the text and tables and was also responsible for formatting the whole to meet the publisher's demanding specification for the preparation of camera ready copy. The high level of technical proficiency and competence with which she undertook these tasks could not have been surpassed.

Abbreviations and Acronyms

ABI	Association of British Insurers
A-i	All-in
APSAB	Average PSA Building
BCCI	Brick construction cost index
BCI	Building cost index
BCIS	Building Cost Information Service
BIR	Board of Inland Revenue
BMI	Building Maintenance Information Limited
BRE	Building Research Establishment
BRS	Building Research Station
CCA	Current cost accounting
CFCCI	Concrete framed construction cost index
CIOB	Chartered Institute of Building
CNC	'Cost of New Construction' (index)
CSO	Central Statistical Office
DERV	Diesel engined road vehicle
DHSS	Department of Health and Social Security
D L & E	Davis, Langdon and Everest
DOE	Department of the Environment
ESCI	Electrical services cost index
FCI	Factor cost index
GBC	General building cost
GBCI	General building cost index
GDP	Gross domestic product
HCI	Housing cost index
HMSO	Her Majesty's Stationery Office
HTPI	Housing tender price index
H & V	Heating and ventilating
IBTP	Index of building tender prices
M & E	Mechanical and electrical
MIPS	Median index of public sector building tender prices
MSCI	Mechanical services cost index
NEDO	National Economic Development Office
OPI	Output price index
PAFA	Price adjustment formula application
PC	Prime cost
PILAH	Price index of local authority housebuilding
PIPSH	Price index of pubic sector housebuilding
PSA	Property Services Agency

PSBTPI	Public sector building tender price index (DOE)
QSSD	Quantity Surveyors Services Division
RCTPI	Road construction tender price index
RPI	Retail prices index
SDD	Scottish Development Department
SFCCI	Steel framed construction cost index
SO	Scottish Office
TPI	Tender price index

Introduction

The fundamental purpose of this book is to provide help to all those who require information about changes in construction costs and prices over time by allowing them to locate an appropriate index and to apply it correctly. We therefore provide a guide to the nature and sources of construction indices and a comprehensive collection of the data available. The number of indices available is now very large; indeed their preparation over recent years has been something of a growth industry. In all, more than 90 series are included here. This large number provides one of the main justifications for the book because the sheer volume of data presents the potential user with considerable difficulties in locating the information he or she needs and making a suitable choice among a variety of alternatives. Many pitfalls confront the ignorant and the unwary.

It is clear that some guidance as to the use and interpretation of the available data is needed. The first part of the book, therefore, is devoted to matters which are important in this context. Chapter 1 discusses the wide range of uses for which cost and price indices are required in construction and considers their practical application in detail. Chapter 2 examines the problems that confront reliable measurement in this field and the methods of measurement that are used. These chapters provide the essential background for an understanding of the nature of the indices available and their use and interpretation. The rest of the book provides full background information about each of the available indices and a complete set of data.

In the rest of this introductory chapter we first provide an introduction to the general concept and use of index numbers and then explain the organisation and content of the book in detail.

The concept of an index

Index numbers of costs and prices provide a convenient means of expressing changes over time in the costs or prices of a *group* of related products in a single summary measure. Most people are familiar, for example, with the official monthly index of retail prices (RPI), or 'cost of living' index, which measures the average change each month in the retail prices of the wide range of goods and services purchased by the average household - often referred to as the 'rate of inflation'. Index numbers are now widely used, not only for measuring price changes but also for a variety of other purposes in many fields where a single summary measure of change is required for a range of different items. A brief account of their historical development is given in Tysoe (1981), pp.1-2.

The concept and use of an index is best explained further by way of an example. There are many ways of expressing the change in cost of a particular item over time. As an illustration of this consider the hourly labour rates from 1980 to 1990 shown below.

Alternative methods of expressing a change

Year (1)	Hourly rate (£) (2)	Absolute change from one year to the next (£) (3)	Percentage change per year (4)	Index (5)
1980	1.48	-	-	100
1981	1.63	0.15	10.1	110.1
1982	1.75	0.12	7.4	118.2
1983	1.85	0.10	5.7	125.0
1984	1.94	0.09	4.9	131.1
1985	2.04	0.10	5.2	137.8
1986	2.16	0.12	5.9	145.9
1987	2.27	0.11	5.1	153.4
1988	2.42	0.15	6.6	163.5
1989	2.63	0.21	8.7	177.7
1990	2.88	0.25	9.5	194.6

The disadvantage of expressing change in absolute terms (column 3) is that the significance of that change is not apparent. In this example it is only apparent when compared with the hourly rate. For example, the increase of £0.15 per hour between 1980 and 1981 and also between 1987 and 1988 was more significant in 1981 than in 1988. In 1981 the £0.15 increase was on an existing rate of £1.48 per hour which was a change of 10.1 per cent. In 1988 the £0.15 increase was on an existing rate of £2.27 per hour, which was a change of 6.6 per cent. This is considerably less, although the absolute change was the same.

This disadvantage is partly overcome by the use of a percentage (column 4) although there is still a drawback in that it is difficult to compare this change other than between consecutive years. The way in which difficulties such as these are overcome is by expressing the change in terms of an index (column 5). In this example, the first year of the series has been designated as the base year when the hourly rate was £1.48. In many comparisons 100 is used as a base, percentages being the most obvious example. In consequence, people are used to such comparisons and statisticians take advantage of this fact by basing index numbers on 100. The method of calculating index figures in this manner is as follows:

Base year index (1980) $= \dfrac{1.48}{1.48} \times 100 = 100$

Index for 1981 = $\dfrac{\text{Figure for 1981}}{\text{Figure for 1980}} \times 100$

= $\dfrac{1.63}{1.48} \times 100 = 110.1$

In mathematical terms the index for any year can be stated as follows:

Price index = $\dfrac{P_1}{P_0} \times 100$

where P_0 is the price in the base year and P_1 is the price in the year under review.

Thus, in summary, the index numbers express the value for the year under review as a percentage of the base year. In the example here, the index number for 1990, 194.6, shows that the hourly labour rate increased by 94.6 per cent from 1980 to 1990 or, in other words, that the rate in 1990 was 1.946 times the 1980 rate.

This example is a very simple one involving a *single* item of interest. The main benefit of index numbers, however, lies in the summarisation of changes for more than one item. We turn to this aspect after giving an illustration of how the index shown above might be used. Consider the following example:

Example

If a contractor's total wage bill in 1981 was £145,000, how much would it be in terms of 1984 labour rates?

The formula can be applied as follows:

(a)   £145,000   x $\dfrac{131.1}{110.1}$ =   £172,657

where 131.1 is the 1984 index
       110.1 is the 1981 index.

or

(b) $\dfrac{131.1 - 110.1}{110.1}$ =   0.19074 increase

        £145,000 x 0.19074   =   £27,657 increase

              total wage bill: £27,657 + £145,000 = £172,657

The actual increase is shown in example (b) which can be expressed in the formula: the increase equals the old cost figure times the difference between the new and the old index all divided by the old index.

increase = $\dfrac{\text{old cost figure (new index - old index)}}{\text{old index}}$

Construction of an index

The index outlined above was for one item involved in comparisons between different periods and the calculation of the index numbers was very simple. Unfortunately, index numbers are required in circumstances where there is more than just one item. In these circumstances each item has to be 'weighted' to allow for their relative importance. The example below illustrates the principle of 'weighting' to calculate a weighted average of hourly earnings for manual and non-manual workers. Columns (2) and (3) in the table below show the average hourly earnings for manual workers (M) and for non-manual workers (N). To calculate an overall average, allowing for the relative importance of these two types of labour, it is necessary to know what weight to attach to each type. In this example we assume that for every two manual workers there is one non-manual worker. Columns (4) and (5) then give weighted totals for each type of labour (that is 2M and 1N respectively) and column (6) gives the sum of the two weighted items, i.e. 2M + 1N. The weighted average - given in column (7) - is simply the weighted total divided by the total weights, i.e. (2M + 1N)/(2 + 1).

Calculation of weighted average and weighted index of earnings

Year	Hourly Earnings		Weighted Earnings				Index
	Manual (£)	Non-manual (£)	Manual (£)	Non-manual (£)	Weighted Total (£)	Weighted Average (£)	
(1)	M (2)	N (3)	2M (4)	1N (5)	2M+1N (6)	(2M+1N)/3 (7)	(8)
1985	3.46	5.14	6.92	5.14	12.06	4.02	100
1986	3.68	5.58	7.36	5.58	12.94	4.31	107.2
1987	3.97	6.06	7.94	6.06	14.00	4.67	116.2
1988	4.22	6.83	8.44	6.83	15.27	5.09	126.6
1989	4.56	7.74	9.12	7.74	16.86	5.62	139.8

Having now obtained a single series of weighted average earnings, the calculation of an index follows the same principles as in the earlier example, that is to say, the weighted average figure for each year is expressed as a percentage of the corresponding figure for the base year. The results are given in the final column.

The index shows that hourly earnings, for all workers as a group, increased overall by 39.8 per cent over the period from 1985 to 1989.

The principles of the calculation shown above may be expressed in general terms as follows:

$$I_t = \frac{[(M_t \times W_m) + (N_t \times W_n)]/(W_m + W_n)}{[(M_o \times W_m) + (N_o \times W_n)]/(W_m + W_n)} \times 100$$

where: I_t denotes the index for the period under review
M_t denotes manual earnings in the period under review
W_m denotes the weight for manual workers
N_t denotes non-manual earnings in the period under review
W_n denotes the weight for non-manual workers
M_o denotes manual earnings in the base period
N_o denotes non-manual earnings in the base period.

Thus the calculation of the index for 1989 is:

$$\frac{[(£4.56 \times 2) + (£7.74 \times 1)]/(2 + 1)}{[(£3.46 \times 2) + (£5.14 \times 1)]/(2 + 1)} \times 100$$

$$= \frac{£5.62}{£4.02} \times 100 = 139.8$$

In perfectly general terms, if we let P_i denote a *set* of prices for different items, the expression for a price index number calculated as a weighted aggregate of several items may then be written as:

$$I_t = \frac{\Sigma P_{it} W_i}{\Sigma P_{io} W_i} \times 100$$

where the subscripts *t* and *o* denote the current period under review and the base period respectively. W_i denotes the set of weights appropriate to each of the items in the set, and the symbol Σ (read as 'sigma') denotes summation, i.e. in this case the summation of the product of P_i and W_i across all items in each period.

A final technical point for those for whom the mechanics and terminology of index number construction is unfamiliar, concerns the terminology which is commonly employed. Some indices are referred to as being Laspeyres indices or Paasche indices. Laspeyres indices involve the use of weights appropriate to the base period - 'base weights' - while Paasche indices involve the use of weights appropriate to the current period - 'current weights'. These indices may therefore be expressed as:

Laspeyres (weighted aggregate) index

$$\frac{\Sigma P_{it} W_o}{\Sigma P_{io} W_o} \times 100$$

Paasche (weighted aggregate) index

$$\frac{\Sigma P_{it} W_t}{\Sigma P_{io} W_t} \times 100$$

In practice most indices are of the Laspeyres type, but the UK Index of Retail Prices (RPI), for example, is a Paasche index. A more detailed, but brief, account of the principles of index number construction will be found in Fleming and Nellis (forthcoming 1991).

When constructing an index, four factors need to be considered. These are as follows:

(a) The purpose of the index.
(b) The selection of the items.
(c) The choice of weights.
(d) The choice of the base year.

(a) The purpose of the index
This must be carefully considered because it will affect decisions relating to the other three factors. Moreover, the interpretation of the index will also depend on the purpose. For example, an index constructed to measure change in factory building costs must not be used for revaluing the plant and machinery, nor even the commercial value of a building, since such an index would not take into account changes such as the value of land on which such buildings are situated.

(b) The selection of the items
This can be the most difficult problem of all. In considering the construction of a cost of living index, obviously bread should be included, but should table wines be included? Heating costs must be included but should television running costs be included? If home rentals are included should holiday rentals be included? The correct solution lies in defining the purpose of the index carefully and then selecting the items that best achieve that purpose.

(c) The choice of weights
It is important to select weights which reflect the relative importance of the constituent items to be included in an index. For example, a set of weights reflecting the relative importance of different inputs into housebuilding would not be appropriate for an index measuring the building costs of, say, warehouses.

(d) The choice of the base year
The base year should be a recent year when there had been no unusual occurrences. The choice of a freak year is a favourite ploy of those who use

statistics to mislead. A dishonest capitalist could choose a record year for profits as base and so show subsequent profits to be painfully low whereas a dishonest trade unionist could similarly choose a year of exceptionally full employment to show that current unemployment is intolerably high.

Basic types of indices used in the construction industry
There are three main kinds of indices used in the construction industry:

(a)   Building cost indices.
(b)   Tender price indices.
(c)   Output price indices.

The terms 'tender prices' and 'building costs' are often confused when considering building indices. 'Building costs' are the costs actually incurred by the builder in the course of his business, i.e. wages, material prices, plant costs, rates, rents, overheads and taxes etc. 'Tender prices' represent the cost a client must pay for a building. They include building costs but also take into account market considerations and thus allow for profits and the builders' anticipation of cost changes during the lifetime of the contract. This means that, for example, in times of boom tender prices may increase at a greater rate than building costs, whilst in a depression the opposite may apply. Output price indices are derived from tender price indices for a special purpose. They are used as deflators to convert official statistics of contractors' output of new construction work from current prices to constant prices. The methods used to compile these indices are given in Chapter 4.

The organisation of the book
The existence of a large number of cost and price indices for construction, as indicated earlier, is a reflection of four factors. Firstly, the diversity of construction work means that different types of work may require their own index. Secondly, indices are required for different purposes: updating tenders, measurement of changes in contractors' costs, revaluing output at constant prices for statistical purposes and a variety of other uses; these are discussed fully in chapter 1. Thirdly, many difficulties confront measurement in a field so diverse as construction where there is no standard product to form the basis of cost measurement on a fully comparable basis over time. These difficulties have bred a variety of attempted solutions. Fourthly, a number of series cover different time periods.

The simplest way of helping the user is to provide a complete collection of data, and this we do. However, one problem facing the user of long-run series is that breaks frequently occur in the series due to periodic changes in the base date of the index. Where this occurs we have calculated a continuous series by rescaling the series arithmetically. It should be recognised that where the rebasing was due to a revision in the method of calculation of the index (for instance the adoption of new weights) this procedure is not statistically correct. However, it is common practice and provided the revisions have not constituted fundamental changes in data or methodology and exceptionally long time-periods are not involved, then the errors introduced by this procedure are not likely to be so large as to invalidate the results. The rescaling we have carried out is clearly indicated in footnotes to the tables. When rescaling is carried out here, the base date chosen is always the one currently in use. This makes it convenient to update the index by simply appending the latest figures and space has been left in the tables for this purpose. To facilitate updating, full details of sources including publication details are given and names, addresses and telephone numbers of the responsible organisations are listed in Appendix B.

Each index is presented in both tabular and graphical form. It should be noted that all graphs are presented using a logarithmic scale for the index numbers on the vertical axis. The aim is to aid interpretation of trends over time because equivalent rates of change are represented by lines having identical slopes when plotted on a logarithmic scale. The example below illustrates this property. The following figures show a series of index numbers over 15 time periods growing by a constant 10 per cent in each successive time period.

Time Period	Index
0	100.00
1	110.00
2	121.00
3	133.10
4	146.41
5	161.05
6	177.16
7	194.87
8	214.36
9	235.80
10	259.37
11	285.31
12	313.84
13	345.23
14	379.75
15	417.73

When these are plotted on an ordinary - non-logarithmic scale - it appears that the rate of increase in the series is accelerating over time - see Figure 1 (Part a). When plotted on a logarithmic scale, however, the graph appears as a straight line and shows, therefore, that the rate of increase is in fact constant - see Figure 1 (Part b). In any one graph, therefore, equal slopes denote equal rates of change.

The choice of an appropriate series in practical applications, requires an appreciation of the different types of construction index that may be defined and an understanding of the way the index has been calculated. Part B of the book presents all of the currently-compiled indices, including background information about methods of calculation, sources of data etc. In this part, particular attention is devoted to the systematic classification of the available indices and the adoption of a standard format for describing each index. All

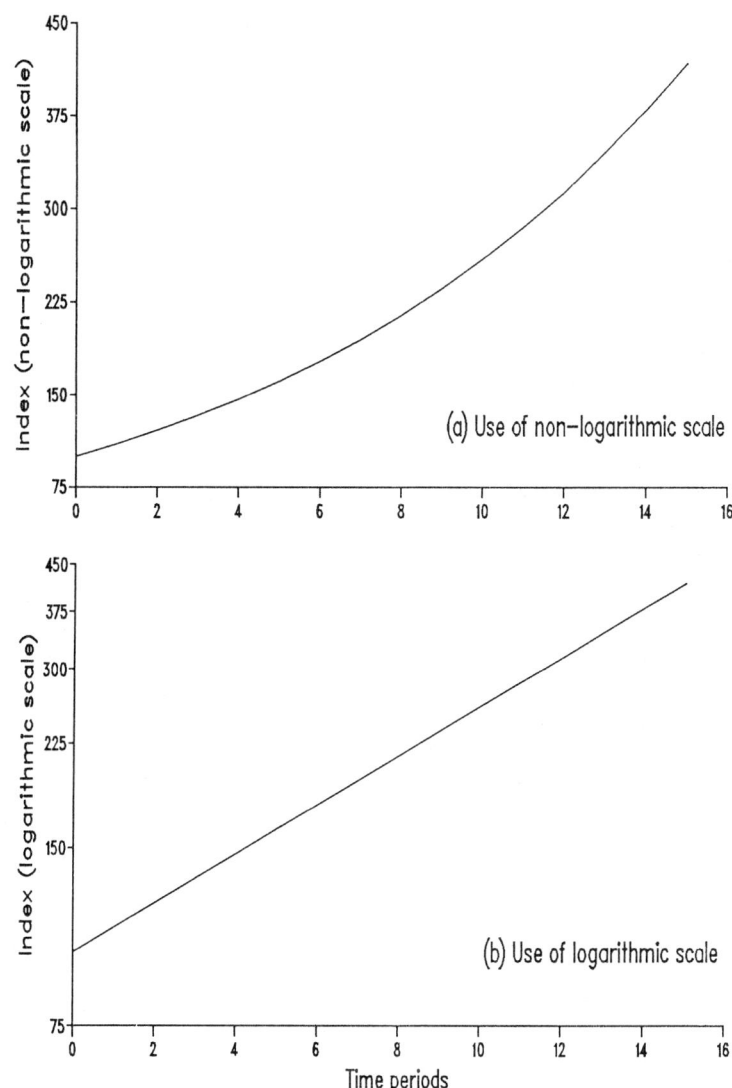

Figure 1 Comparison of Logarithmic and Non-logarithmic Scales

of the indices are classified by type and considered in separate chapters as follows:

(a) Output price indices - Chapter 4.
(b) Tender price indices - Chapter 5.
(c) Cost indices - Chapter 6.

The background information about each index is presented following a standard format as follows:

Type of index
Series: coverage and breakdowns
Base dates and period covered
Frequency
Geographical coverage
Type and source of data
Method of compilation
Commentary
Publications

Part C of the book presents historical series. These cover various periods of time extending back to the first half of the nineteenth century. These are not classified according to type in the same way as the current indices because in some cases the method of calculation is not known and in other cases because the method is based on a composite of cost and price information. The importance of long-run historical indices rests upon the fact that buildings and other construction works have long lengths of life and, as a consequence, correspondingly long time series measuring changes in construction costs and prices are required for some purposes. In the main, the uses in this area are more important to statisticians and economists concerned with the compilation of output and capital stock statistics and studying investment, economic growth and inflation. But practitioners concerned with the revaluation of construction assets also have a call for long-run indices.

Finally, it is of interest to compare movements in construction costs and prices over time with the corresponding movements of costs and prices in the economy as a whole. Indices relevant in this context are presented in Appendix A as follows:

1. Cost of living index, 1914-1948
2. Index of retail prices, 1948-1989
3. Index of total home costs, 1948-1989
4. Index of capital goods prices, 1914-1989

Comparisons between the trends revealed by these indices and those revealed by representative indices for construction cost and price movements are made in summary chapters which conclude Part B (chapter 7) and Part C (chapter 10).

Part A

Construction Indices: Uses and Methodology

1

Uses of Construction Indices

Indices are used for many purposes. This chapter explains most of the uses to which construction indices are put with an emphasis on the uses of tender price indices.

The uses covered in this chapter are:

Assessing the level of individual tenders
Adjustment for time - updating and backdating
Pricing
Cost planning
Forecasting
General comparisons
Comparisons between published indices - changing the base date
Variation of price clauses
Calculation of cash-flow projections
Calculation of derived tender price indices

The uses of indices covered in this chapter assume that the reader has selected the most appropriate index. In making his choice the reader should have a knowledge of the construction of the selected index. He should be aware of the type of index that he is using and be sure that he is using it for the purpose for which it was intended. Care must be taken in the application of the index. He should be sure that the answer is reasonable and that there are no underlying factors, either within the use that is being made of the index or within the index itself, which invalidate the answer.

The user must not fall into the trap that so many have fallen into in the past and use a factor cost index to update a tender for any of the purposes described in this chapter. This is the most common misuse that seems to be made of construction indices, and one that should be avoided. The best advice that can be given to the reader about the selection of the most appropriate index is for him to understand the compilation and purpose of any index that he intends to use.

1.1 ASSESSING THE LEVEL OF INDIVIDUAL TENDERS

The level of individual tender prices may be assessed, in order to judge their competitiveness, by 'indexing' the tender in the way described below and comparing it with a benchmark given by one of the general tender price indices (considered in chapter 5). Tender price indices are generally based on a sample of tenders in each time period. Each of these tenders is 'indexed', that is to say the tender is revalued (in whole or in part) using a base schedule of rates and the current value expressed as a percentage of the rebased value. This gives an individual 'project index'. The average (either arithmetic or geometric mean or median) of the project indices constitutes the general index of tender prices. This methodology is explained more fully later (see chapter 5). Its relevance in the present context is that it is possible to assess the level of an individual tender as a measure of its competitiveness, by 'indexing' the tender, in the way described above, and comparing its level with that of the general tender price index: the lower it is relative to the general index then the 'keener' it may be said to be and *vice versa*. This information will have an important bearing on any conclusion reached from the analysis of the result of a tender. Any cost planning carried out prior to tendering can be viewed with regard to this information. The level of a tender is then another yardstick by which it can be judged.

There are two scales that ought to be considered together in an analysis of the results of a tender. One is the monetary result when compared to the budget/cost limit, and the other is the relationship between the index of the tender and the norm for the period. These two should not be considered in isolation. Should the tender be at a relatively similar position on both scales (i.e. either at the top, bottom or points in between) it would follow that the effect on the monetary scale is more than likely, although not of necessity, the result of the other. Should the tender not be similarly positioned on both scales, and the effect on the monetary scale is not attributable to the tendering climate, the cause can be looked for elsewhere. The cause could be inadequate cost planning, if the monetary level is in excess of the budget/cost limit whilst the tender index indicates that the tender is keen; or that the budget/cost limit was unrealistic; or a combination of both.

1.2 ADJUSTMENT FOR TIME

When it is required to adjust a tender, cost analysis, rate or the like because they are required to relate to a different time period from that for which they were originally compiled, then a suitable index should be used to make this adjustment. The method of adjustment is given below.

If the cost to the client is to be measured when adjusting a tender, cost analysis, rate or the like then a tender price index should be used. It will obviously be more accurate if the index figure for the individual tender etc. is used rather than the mean index figure for the period.

The calculation for adjusting a tender, cost analysis, rate or the like in order that it may relate to a different time period is as follows:

$$A = \frac{(B - C)\,100}{C}$$

where A = percentage change
B = index number at the date to which the tender etc. is being adjusted
C = index number of the tender etc. or index number current at date of tender

Examples

1. Date of tender: May 1986.

It is required to adjust the tender to relate to February 1989. From the DOE public sector building tender price index (firm price) (1985 = 100 base) (see Chapter 5, section 5.1):

$$B = 136 \text{ and } C = 102$$

$$\text{Thus } A = \frac{(136 - 102)\ 100}{102} = 33.333\%$$

The 'instant' way of obtaining the above percentage, of course, is to divide 136 by 102 (= 1.33333) move the decimal point two places to the right and deduct 100. The multiplier to use for updating the tender etc. in question would be 1.33333 (136 divided by 102).

2. Date of tender: February 1989

It is required to adjust the tender to relate to May 1986. From the DOE public sector building tender price index (firm price) (1985 = 100 base) (see Chapter 5, section 5.1).

$$B = 102 \text{ and } C = 136$$

$$A = \frac{(102 - 136)\ 100}{136} = -25\%$$

The percentage is negative, showing that it has to be deducted. The multiplier to use for backdating the tender etc. in question would be 0.75 (102 divided by 136).

1.3 PRICING

Pricing a Bill of Quantities can be done in many ways. One method is to price the scheme at the rates used in another project of similar type in a similar location and adjust by means of indices for the difference in time. As any project could have a range of level of pricing of 30 per cent about the mean, it follows that it is important to know the individual index value of any project which is to be used as a basis for pricing some later scheme.

When using an index to facilitate pricing, care must be taken regarding preliminaries etc. The example that follows assumes that the preliminaries etc. of the tender being used and the scheme in question are at the same level.

16   *Spon's handbook of construction cost and price indices*

Example

Index of tender being used: 106
Assumed current level of tendering: 133
Price being used - rate per cubic metre of plain
    concrete (1:2:4) in foundations in trenches   £49.09
To update:

$$A = \frac{(B - C) \, 100}{C}$$

Where A = required percentage to be added
      B = 133
      C = 106

$$A = \frac{27 \times 100}{106} = 25.47\%$$

The rate to be used in the new scheme is therefore:

£49.09 + 25.47% = £61.59 per cubic metre.

This updated rate assumes that the preliminaries etc. in both schemes are at the same level. Should this not be the case then a suitable adjustment would have to be made. The following formula would be used:

$$R2 + Y = R1 + X$$

where R1 is the updated rate unadjusted for preliminaries etc., R2 is the updated rate adjusted for preliminaries etc., X are the preliminaries etc. of the tender being used, as a percentage, and Y are the preliminaries etc. of the new scheme, as a percentage.
    Assuming X = 10% (i.e. one tenth) and Y = 20% (i.e. one fifth) the adjustment would be as follows:

$$R2 + \frac{R2}{5} = 61.59 + \frac{61.59}{10}$$

$$\frac{6R2}{5} = 67.749$$

$$R2 = \frac{5 \times 67.749}{6} = £56.46$$

1.4 COST PLANNING

In the process of cost planning, cost information that is available concerning previous schemes will be out of date and will have to be updated before use. If, for example, in the early stages of cost planning an office block, the cost analyses of similar buildings were available together with their individual indices, a rate per square metre estimate of the likely cost could be arrived at. The original building cost analysis could be updated and the resultant rate per square metre applied to the area of the proposed new building.

Example

Total cost analysis figure	£1,704,000.00
Total area	2400 square metres
Rate per square metre of cost analysis	£710.00
Total area of proposed new building	2500 square metres
Index of original scheme	106
Current index	133

To update:

$$A = \frac{(B - C) \, 100}{C}$$

where A = required percentage to be added
B = 133
C = 106

$$A = \frac{27 \times 100}{106} = 25.47\%$$

Rate to be applied to area of proposed new building is therefore:

£710.00 + 25.47% = £890.84 per square metre
2500 square metres at £890.84 = £2,227,100.00

Therefore the approximate estimate of the building cost of the proposed new office block is £2,277,100.00.

1.5 FORECASTING

Indices can play a major role in forecasting cost trends to both the client and to the building contractor. The results of such forecasting can be used as a basis for the calculation of the following:
(a) The value of the return to the contractor under the fluctuations clause of the contract (see p.20).
(b) The cost to the client of a new building at some future time (see below).
(c) The likely monthly payments during the course of a building contract (see p.21).

The BCIS produce a forecast for fluctuations for forecasting reimbursement to the contractor under the fluctuations clause of the contract. Future changes in factor costs (see Glossary) are predicted. From these predictions the forecast net amount recoverable can be calculated. Their latest forecast is available on subscription.

18   *Spon's handbook of construction cost and price indices*

An index of building costs appears annually in *Spon's Architects' and Builders' Price Book*. The authors also forecast future changes in factor costs (see Glossary) to produce a projection of that index. They take this a step further by predicting the likely effects of the tender climate on these changes when compared with their tender price index. These appear in the magazine *Building* from time to time in both graphical and tabulated form.

Once produced, such forecasting indices can be used to estimate various costs in the same manner that historical and current indices are used. The last example used under cost planning could be employed here, the only change being a change in terminology, i.e. for current index read predicted index.

1.6 GENERAL COMPARISONS

The availability of data limits the extent to which indices can be used for comparative purposes.

Example

As an example, consider a hypothetical case, assuming that the number of tenders in the sample of a tender-based index for a quarter was 160, giving a mean index of 220. Assume also that the sample was composed of 80 housing schemes having a mean index of 225 and 80 health buildings having a mean index of 215. In addition, assume that for a later quarter, the number of tenders in the sample was again 160 with a mean index of 270. Assume that this sample consisted of 80 housing schemes having a mean index of 290 and 80 health buildings having a mean index of 250. The comparisons between these two quarters could be tabulated as follows:

	Published index	Housing content	Health building content
Earlier quarter	220	225	215
Later quarter	270	290	250
Increase of later quarter over earlier quarter	22.73%	28.89%	16.28%

The substantial difference in the increase for the housing and health constituents of the published index indicate that more valid conclusions can be determined when additional data for specific purposes - as for specific types of building - is given. The BCIS publish a Building Cost Index for four classes of building (see Chapter 6, section 6.1). This enables their relative movement to be assessed.

1.7 COMPARISONS BETWEEN PUBLISHED INDICES

The comparison of two indices is sometimes hindered by the fact that their base dates differ. For example, Table 1.7.1 below reproduces two of the tender price indices available: the BCIS tender price index on a firm price basis and the D L & E tender price index (see Chapter 5, sections 5.3 and 5.7) from 1986(1) to 1988(2).

Table 1.7.1

Year	Qtr	BCIS tender price index firm price 1985 = 100	D L & E tender price index 1976 = 100
1986	(1)	100	221
	(2)	102	226
	(3)	104	234
	(4)	105	234
1987	(1)	107	242
	(2)	105	249
	(3)	108	265
	(4)	116	279
1988	(1)	118	289
	(2)	121	299

Source: BCIS Manual and *Spon's Architects' and Builders' Price Book*.

For the period in Table 1.7.1 it can be seen that the BCIS tender price index has increased by 21 per cent {[(121 - 100)/100] x 100} and that the D L & E tender price index has increased by 35 per cent {[(299 - 221)/221] x 100}. It can also be seen that there have been different peaks and troughs. It is difficult to compare them properly, however, because they do not have the same base. It is, therefore, desirable to rescale one of the series to equate its base to that of the other series. The D L & E tender price index is changed to a base of 1985 = 100 by dividing each index number by the index at the required new base date, multiplied by 100. The formula to be applied is as follows:

$$A = \frac{B \times 100}{C}$$

where A is the new index number with the new base date; B is the old index number with the original base date; and C is the old index number of the new base date. In this case C = 218; this is the value of the D L & E index in 1985 with base 1976 = 100 (see Table 5.7 in Chapter 5).

If this formula is applied to the indices shown in Table 1.7.1 it can be reproduced in the form of Table 1.7.2 below.

Table 1.7.2

Year	Qtr	BCIS tender price index firm price 1985 = 100	D L & E tender price index 1985 = 100
1986	(1)	100	101
	(2)	102	104
	(3)	104	107
	(4)	105	107
1987	(1)	107	111
	(2)	105	114
	(3)	108	121
	(4)	116	128
1988	(1)	118	132
	(2)	121	137

Troughs and peaks can now be compared with ease and, bearing in mind the input, conclusions can be reached. The D L & E tender price index 1988(2) figure is well in advance of the BCIS tender price index 1988(2) figure. This is because the BCIS tender price index in question is purely for firm price tenders, whereas the D L & E tender price index is purely for fluctuating price tenders in Greater London. Prices for these types of tenders in this specific location clearly rose at a faster rate than those for firm price tenders nationally.

Indices are often compared to see if there are any differences, and if there are, to consider the possible reasons for such differences. If any differences can be logically explained, one's confidence in the indices being compared is increased.

1.8 VARIATION OF PRICE CLAUSES

A number of indices are published which are used as a basis for the payment of increases or decreases in cost to or from the contractor under the variation of price clauses of the contract.

The most widely used set of indices of this kind are used in the PSA (formerly NEDO) formula (see Chapter 6, section 6.8). These indices are contained in *Price Adjustment Formulae for Construction Contracts: Monthly Bulletin of Indices*, published by HMSO. It has been concluded as a result of a survey by NEDO, that there are both advantages and disadvantages from the use of a formula method using indices as a basis for the calculation of the payment of increases or decreases in cost to or from contractors, as opposed to a conventional method, as follows:

(a) The advantages of the conventional method are that it is simple in principle and capable of detailed audit.
(b) The disadvantages of the conventional method are that it is costly to prepare and to check; that market price is difficult to define; that there is

a shortfall (assessed by NEDO at between 25 and 40 per cent) and that there is delay in receipt which can adversely affect cash flow.
(c) The advantages of the formula method are that administrative costs to the contractor are reduced; that government statistics are used; that it is more equitable and that there is less delay in reimbursement.
(d) The disadvantage of the formula method is that there is no exact reimbursement of the contractors' increased costs.

1.9 CALCULATION OF CASH-FLOW PROJECTIONS

Once a building contract has been let it is usual for the quantity surveyor to advise the client of his likely monthly financial commitments. The Department of Health has produced an effective method of forecasting the expected monthly valuations from the contract data. This method was explained in detail in an article entitled 'DHSS expenditure forecasting method' by K.W. Hudson, FRICS in the *Chartered Surveyor Building and Quantity Surveying Quarterly*, 5(3), Spring 1978.

Once the forecast monthly valuations have been obtained, the payment of increases or decreases in cost to, or from, the contractor, under the variation of price clauses of the contract, can be calculated from a forecast projection of a factor cost index, in the same way as described in Chapter 6, section 6.8 under the PSA formula. This gives a better service to clients and permits a closer monitoring of the contractors' performance.

Example

Assume that a tender for a new office block has been received totalling £900,000 and that the contract period is 18 months. Assume that the base month under the fluctuations clauses of the contract, which is on a formula basis, has an index of 100 on a factor cost index, and that there is no non-adjustable element. Assume that the month by month projection of that factor cost index over the contract period is as follows:

Month	Factor cost index	Month	Factor cost index
1	103	10	115
2	104	11	117
3	105	12	119
4	106	13	120
5	108	14	120
6	109	15	125
7	111	16	126
8	113	17	128
9	113	18	130

22   *Spon's handbook of construction cost and price indices*

The standard cash flow from the DHSS method of forecasting would be as follows for this scheme:

Month	£	Month	£
1	21,819	10	70,202
2	33,377	11	67,815
3	43,388	12	63,955
4	51,855	13	58,513
5	58,776	14	51,526
6	64,152	15	42,994
7	67,983	16	32,916
8	70,267	17	21,293
9	71,008	18	8,125

Using the formula, further explained in Chapter 6, section 6.8, under the PSA formula, the payment of increases in cost to the contractor under the variation of price clause can be calculated as follows:

$$C = V \times \frac{I_v - I_o}{I_o}$$

where C is the amount of the price adjustment to be paid to the contractor,
V is the forecast value of work to be executed during the valuation period,
I_v is the index number current at the mid-point of the valuation period (for this example this has been taken as the projected monthly index that coincides with the valuation month),
I_o is the index number for the base month.

Month 1
$$C = £21,819 \times \frac{(103 - 100)}{100} = £655$$

Month 2
$$C = £33,377 \times \frac{(104 - 100)}{100} = £1,355$$

Month 3
$$C = £43,388 \times \frac{(105 - 100)}{100} = £2,169$$

Month 4
$$C = £51,855 \times \frac{(106 - 100)}{100} = £3,111$$

Uses of construction indices 23

Month 5
$$C = £58,776 \times \frac{(108 - 100)}{100} = £4,702$$

Month 6
$$C = £64,152 \times \frac{(109 - 100)}{100} = £5,774$$

Month 7
$$C = £67,983 \times \frac{(111 - 100)}{100} = £7,478$$

Month 8
$$C = £70,267 \times \frac{(113 - 100)}{100} = £9,135$$

Month 9
$$C = £71,008 \times \frac{(113 - 100)}{100} = £9,231$$

Month 10
$$C = £70,202 \times \frac{(115 - 100)}{100} = £10,530$$

Month 11
$$C = £67,851 \times \frac{(117 - 100)}{100} = £11,535$$

Month 12
$$C = £63,955 \times \frac{(119 - 100)}{100} = £12,151$$

Month 13
$$C = £58,513 \times \frac{(120 - 100)}{100} = £11,703$$

Month 14
$$C = £51,526 \times \frac{(120 - 100)}{100} = £10,305$$

Month 15
$$C = £42,994 \times \frac{(125 - 100)}{100} = £10,749$$

Month 16
$$C = £32,916 \times \frac{(126 - 100)}{100} = £8,558$$

Month 17
$$C = £21,293 \times \frac{(128 - 100)}{100} = £5,962$$

Month 18
$$C = £8,125 \times \frac{(130 - 100)}{100} = £2,438$$

The cash flow can then be given as the forecast expenditure plus the forecast payment of increases in cost to the contractor as follows:

Month	Forecast cash flow (£)	Month	Forecast cash flow (£)
1	22,474	10	80,732
2	34,712	11	79,386
3	45,557	12	76,106
4	54,966	13	70,216
5	63,478	14	61,831
6	69,926	15	53,743
7	75,461	16	41,474
8	79,402	17	27,255
9	80,239	18	10,563

The above figures do not include any adjustment for retention which would have to be made, depending on the type of contract, nor do they differentiate between the building and engineering elements.

1.10 CALCULATION OF DERIVED TENDER PRICE INDICES

Although it is best to use an index that has been compiled specifically for a given purpose, there are times when a required index is not available.

There is a need within the construction industry for tender price indices for engineering installations. This need has not yet been satisfied despite a number of attempts at compilation by various organisations. In the absence of engineering tender price indices it is a hypothesis of ours that, making the basic assumption that the market conditions in the building industry are similar to those for engineering, it is possible to derive the required indices. They could be derived from the relevant engineering factor cost indices coupled with the tender price indices and the related factor cost indices that are available for building.

Consider the table below, which shows a factor cost index for building together with a related tender price index for building, and a factor cost index for electrical installations. Is it possible to construct a tender price index for electrical installations from these data?

Table 1.10.1

Base: 1985 = 100

		Building		Electrical Installations	
Date		APSAB index for building	DOE public sector tender price index	APSAB index for electrical installations	?
Year	Qtr				
1985	(1)	97	95	104	
	(2)	98	98	105	
	(3)	101	99	107	
	(4)	101	99	109	
1986	(1)	101	101	110	
	(2)	101	101	110	
	(3)	103	98	111	
	(4)	104	104	112	
1987	(1)	104	106	116	
	(2)	105	110	116	
	(3)	108	109	117	
	(4)	108	115	119	
1988	(1)	109	121	124	
	(2)	111	125	124	
	(3)	114	129	124	
	(4)	116	127	126	

Source: PSA, *Quantity Surveyors Information Notes.*

Consider the period from 1987 (4) to 1988 (4). The APSAB index for building has moved from 108 to 116, 7.41 per cent, whereas the DOE public sector tender price index has moved from 115 to 127, 10.43 per cent. Tender prices have therefore moved by 3 per cent more than factor costs {[110.43 − 107.41]/107.41] x 100}.

The APSAB index for electrical installations has moved from 119 to 126, 5.88 per cent. Tender prices for electrical installations ought to have moved by 3 per cent more than their factor costs, if market conditions are assumed to be similar, i.e. 9 per cent {[(1.0588 x 1.03) − 1] x 100}.

The relevant index numbers on the proposed derived electrical installation tender price index for 1987 (4) and 1988 (4) can be found by ratio as follows:

1987 (4): $\dfrac{115 \times 119}{108}$ = 126.7

1988 (4): $\dfrac{127 \times 126}{116}$ = 137.95

26 *Spon's handbook of construction cost and price indices*

The movement from 126.7 to 137.95 represents an increase of 9 per cent from the index numbers, which checks with the percentage addition. The fourth column in the previous table can therefore be filled in as follows:

Table 1.10.2

Base: 1985 = 100

		Building		Electrical Installations	
Date		APSAB index for building	DOE public sector tender price	APSAB index for electrical installations	Derived tender price index for electrical installations
Year		Qtr		index	
1985	(1)	97	95	104	102
	(2)	98	98	105	105
	(3)	101	99	107	105
	(4)	101	99	109	107
1986	(1)	101	101	110	110
	(2)	101	101	110	110
	(3)	103	98	111	106
	(4)	104	104	112	112
1987	(1)	104	106	116	118
	(2)	105	110	116	122
	(3)	108	109	117	118
	(4)	108	115	119	127
1988	(1)	109	121	124	138
	(2)	111	125	124	140
	(3)	114	129	124	140
	(4)	116	127	126	138
		(b)	(a)	(y)	$(x) = \dfrac{(a) \times (y)}{(b)}$

Source: PSA, *Quantity Surveyors Information Notes*, for columns (b), (a) and (y)

In mathematical terms the method can be set out as follows:

Building Indices

Tender price index (TPI) = a_1 in earlier quarter, a_2 in later quarter
APSAB = b_1 in earlier quarter, b_2 in later quarter

$$\text{Ratio of } \frac{\% \text{ increase in TPI}}{\% \text{ increase in APSAB}} = \frac{a_2}{a_1} \times \frac{b_1}{b_2}$$

Engineering Indices

TPI = x_1 in earlier quarter, x_2 in later quarter
APSAB = y_1 in earlier quarter, y_2 in later quarter

$$\text{Ratio of } \frac{\% \text{ increase in TPI}}{\% \text{ increase in APSAB}} = \frac{x_2}{x_1} \times \frac{y_1}{y_2}$$

Assuming the relative changes between factor cost and tender price indices are the same for building and engineering,

$$\frac{x_2}{x_1} \times \frac{y_1}{y_2} = \frac{a_2}{a_1} \times \frac{b_1}{b_2}$$

Therefore

$$\frac{x_2}{x_1} = \frac{a_2\, b_1\, y_2}{a_1\, b_2\, y_1} = \text{factor required to convert an engineering tender figure from earlier to later quarter}$$

In quarterly calculations of values of x, it was assumed that

$$x = \frac{ay}{b}$$

Therefore

$$\frac{x_2}{x_1} = \frac{a_2\, y_2}{b_2} \times \frac{b_1}{a_1\, y_1} = \frac{a_2\, b_1\, y_2}{a_1\, b_2\, y_1}$$

as before.

Any differences between the multiplying factors found from the calculated values of x, and those found from the other three indices directly, will be due to rounding of the calculated x values.

It is our contention that, based on the foregoing, derived tender price indices for engineering installations can be compiled as in the table that follows (Table 1.10.3).

The derived tender price indices must be used with caution. It must be borne in mind that they will only reflect the movement of actual tender prices if the engineering industry, relevant to building, reacts to building market conditions in the same way and at the same time. There can, on occasion, be large differentials, for instance excess activity in the engineering field occasioned by the oil rig production for the North Sea. Although this sort of activity occurs rarely it is usually big enough to be self evident. In the figure that follows (Figure 1.10.3), the derived tender price indices which are a combination of factor costs and market conditions, show a movement from 1985 (1) to 1989 (2) roughly in line with the movement of the tender price index for building. But at other times the indices have diverged. This evidence coincides with the views of engineering colleagues working in these fields.

Table 1.10.3

Base: 1985 = 100

	Building		Electrical Installations		Mechanical Services	
Date Year Qtr	APSAB index for building	DOE public sector building tender price index	APSAB index for electrical installations	Derived tender price index for electrical installations	APSAB index for H&V air con-ditioning	Derived tender price index for mechanical services
1985 (1)	97	95	104	102	97	95
(2)	98	98	105	105	99	99
(3)	101	99	107	105	99	97
(4)	101	99	109	107	99	97
1986 (1)	101	101	110	110	100	100
(2)	101	101	110	110	103	103
(3)	103	98	111	106	103	98
(4)	104	104	112	112	104	104
1987 (1)	104	106	116	118	105	107
(2)	105	110	116	122	108	113
(3)	108	109	117	118	109	110
(4)	108	115	119	127	110	117
1988 (1)	109	121	124	138	111	123
(2)	111	125	124	140	114	128
(3)	114	129	124	140	115	130
(4)	116	127	126	138	117	128
1989 (1)	117	134	131	150	118	135
(2)	120	135	131	147	122	137
Calculation	(b)	(a)	(y)	$(x) = \dfrac{(a) \times (y)}{(b)}$	(d)	$(c) = \dfrac{(a) \times (d)}{(b)}$

Source: PSA, *Quantity Surveyors Information Notes* for columns (b), (a), (y) and (d).

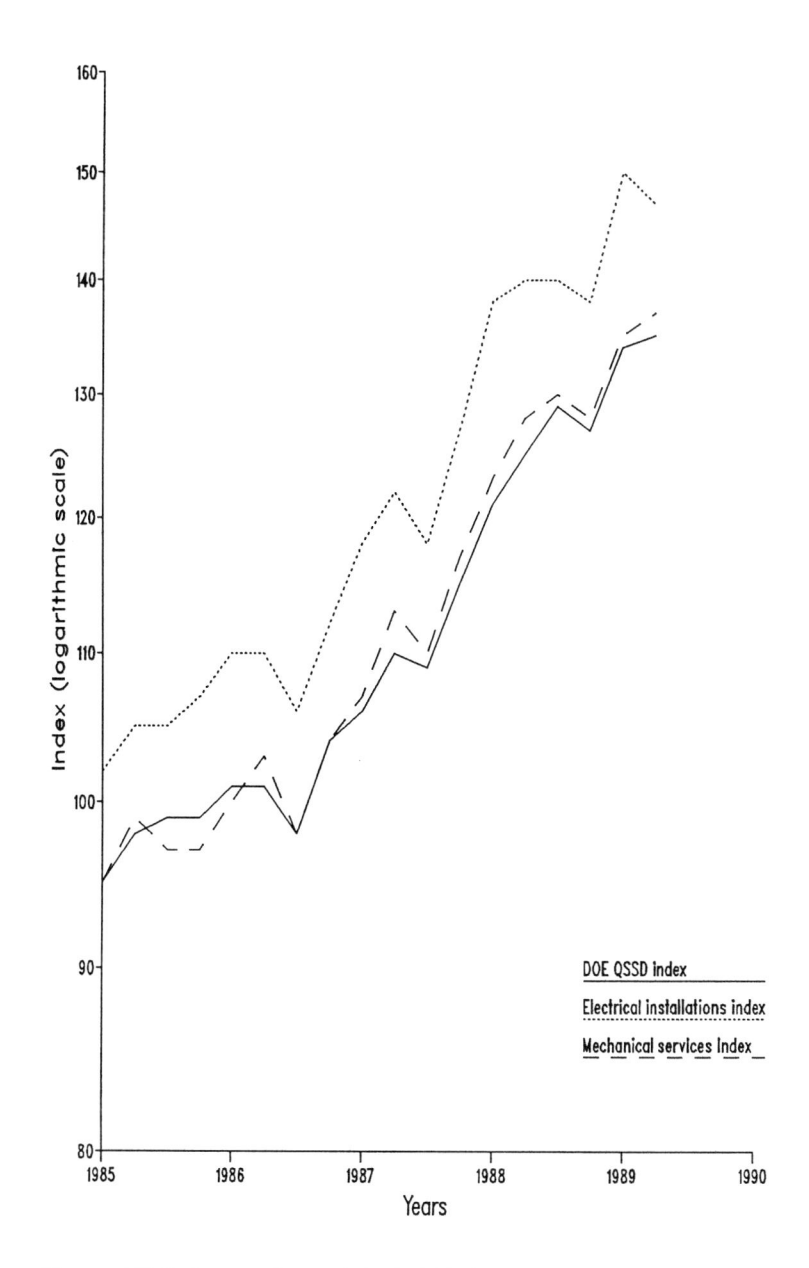

Figure 1.10.3 Comparison of DOE QSSD Public Sector Building Tender Price Index and the Derived Tender Price Indices (1985 = 100), 1985 Q1 – 1989 Q2

2

Problems and Methods of Measurement

This chapter falls into two parts. The first part covers the problems involved in measuring movements in construction costs and prices over time and the second part considers the methods that may be employed. The purpose is to provide the background for understanding the different types of index available and the different methods of calculation that are used.

The distinction between the terms 'costs' and 'prices' has already been explained but it may require some further explanation because they are sometimes used synonymously and at other times not. The term 'construction costs' may be used to refer to the costs incurred by contractors covering labour, materials, plant and possibly overheads but excluding profit. The term 'prices', by contrast, represents the amount paid by contractors' clients and naturally includes profits. But from a client's point of view, the price paid represents his 'costs of construction' and it is in this sense that the two terms may be used synonymously. In this chapter we use the terms interchangeably in this way (unless indicated otherwise) but specific index numbers of costs or prices are clearly differentiated.

The attempt to measure changes in construction costs and prices over time raises a number of conceptual and practical problems. In response to these problems various methods of measurement have been employed and a variety of different series have been devised. Interpretation of the available series must rest upon an appreciation of these problems and the different methods of measurement.

2.1 PROBLEMS OF MEASUREMENT

The basic measurement problem that has to be faced arises from the extremely varied nature of the work carried out by the construction industry. This includes the construction of a wide variety of building and civil engineering works and a wide variety of repair and maintenance jobs. New construction projects vary not only according to type but also in size, design, specification, complexity and methods of construction etc.; even similar jobs vary according to differences in site conditions with a consequential influence upon costs of construction. In some sense, therefore, each job tends to be unique. Thus from the point of view of measuring changes in costs over time there is no single standard of comparison. Another factor to bear in mind is that construction projects often take a considerable period of time from start to finish and that costs may be measured at different stages of the process: they may refer to prices for work yet to be carried out, that is to say, tender prices, or to the level of costs for work currently being executed, or to the costs or prices of work which has been completed - i.e. complete projects ready for use. Clearly, measurements which relate to these three stages may be expected to differ and likewise the rates of change over time of such measurements may not be equivalent.

2.2 METHODS OF MEASUREMENT

We now turn to consider the methods of measurement which may be used. The distinctions we draw are then used throughout the rest of the book as a means of classifying the indices available. Broadly speaking, there are two approaches to the problem of devising an index of construction costs or prices. One is to use price data for actual contracts. The other is to use the information about changes in factor costs, i.e. the costs of the factors of production which go to determine price, namely labour and material costs, overheads and profits after allowing for the influence of productivity changes on prices. Within each of these two broad approaches, certain variants may be defined as follows:

(a)　Actual price data
　　(i)　Total prices
　　(ii)　Unit rates and the repricing of tenders

(b)　Factor costs
　　(i)　Factor cost indices
　　(ii)　Repricing aggregate factor costs
　　(iii) Published unit rates.

We comment on each of these in turn.

(a)　Use of actual price data

Total Prices. Only brief remarks are necessary here because this method is not in current use. The main problem confronting the use of this method is the lack of a standard product so that it is difficult, if not impossible, to make comparisons of building prices on a like-for-like basis covering the whole range of construction outputs. One index, the 'Venning' - later the 'Venning Hope' - index (see Chapter 9, section 9.7) was based informally on information about total tender prices but it ceased publication in 1975. A possible way of overcoming the problem of the lack of a standard product would be to invite tenders periodically from builders for a building of a standard design and specification, even though it was not intended to erect the building. Such a method would be faced, of course, with the need to allow for the changes in standards that do take place over time and with the more severe problem that, since the builders tendering would have no prospect of gaining a contract, there could be no assurance that the prices quoted were reasonable reflections of current cost levels and the tendering climate. The method has not been used in this country. A further approach to the problem of a non-standard product using actual price data is to employ the statistical technique of multivariate regression analysis to devise a statistical model allowing prices to be predicted on the basis of information about the physical characteristics of buildings. Given information about the latter, 'standardised' price comparisons can be made over time. The method is used to compile indices of house prices in the UK (Fleming and Nellis, 1984, 1985, 1989) but has not been used to produce a general construction index.

Unit Rates and the Repricing of Tenders. Instead of using information about the total price of a contract, it is possible to use information about the unit rates for particular categories of work used in building up the total contract price. These are available from the priced bills of quantities of accepted tenders. Such rates refer to specific construction operations and they therefore have the advantage of being directly comparable. Naturally, use of the method requires access to a representative selection of bills and also a reasonably large number because the rates inserted by different builders vary considerably not only because of differences in their levels of efficiency and in the labour, materials and plant costs used by the estimator, but also because of differences in the practices adopted by firms in arriving at a total tender price. Use of the method has been increasingly favoured in Great Britain where, unlike other countries, the use of bills of quantities for tendering purposes provides an extremely valuable data source.

There are basically two ways in which the method may be applied in practice. One is to measure the percentage change in the rates quoted in current tenders compared with base-period tenders (a price relative) and to take an appropriately weighted average of these. This way of using the method is used to obtain a price index for public sector housebuilding (see chapter 5, section 5.5) but not for building costs in general. The other way of applying the method - and one used in several indices of tender prices - is not to compare unit rates directly but to use the rates to re-price tenders. Here, again, two basic approaches are possible: either to use standard rates from a base period to re-price current bills or to use the current rates to re-price a standard bill. Comparison of the values of the bill at base-period and current prices yields an index of price change for each project. These 'project indices' then have to be averaged over a statistically acceptable number of projects to produce a single representative index.

A method analogous to the use of rates extracted from priced bills is to use unit rates published in builders' price books. Such rates, however, are only estimates built up on the basis of certain standardised formulae for combining input costs and as such are more appropriately considered later.

The use of methods based on actual prices has a disadvantage from the point of view of studies concerned with total construction work, in that it will be generally possible to cover only certain well-defined classes of work, the price movements for which may not be representative of all work. Studies concerned with construction work as a whole require a more general measure of price movements. Until a few years ago, such a measure was built up on the basis of changes in factor costs. In 1978 'output price indices' were introduced by the DOE using an alternative methodology which incorporates the information about tender prices referred to above. The methodology is best defined when considering the available series below (Chapter 4). Several series of index numbers based upon factor costs remain, however, as a measure of the movement of construction costs for the contractor.

(b) Use of factor cost data

Weighted Averages of Factor Cost Indices. This method basically consists of taking a weighted average of indices measuring changes in the costs of labour and materials and possibly also contractors' plant. Labour costs may be measured by reference to wage rates possibly with the addition of allowances for labour 'on-costs', such as payments for insurance, training levies, travelling time, payroll taxes etc. The use of wage rates to measure labour costs will not reflect the payment of rates in excess of the basic negotiated rates ('attraction money'), nor payments for overtime, bonuses etc. Official statistics of average earnings in the industry, however, do reflect such payments and these are sometimes used as the basis for an index of labour costs (see *Housing and Construction Statistics,* HMSO). An index on this basis, of course, reflects market conditions. Information on the prices of construction materials is collected and published by various bodies, including the government, and the latter publish indices of housebuilding and construction materials prices (also in *Housing and Construction Statistics,* HMSO). Many factor costs indices rely solely on information on labour and material costs. But official information on plant costs is also published and is incorporated in an index known as the APSAB index (see chapter 6, section 6.8 for further details).

At one time the DOE relied on factor cost information, together with estimates of changes in labour productivity, overheads and profits, to produce indices of the type now referred to as 'output price indices'. It was an attempt to measure changes in the price not of tenders but of work carried out by contractors in each quarter, allowing for the influence of market conditions. It was published as the DOE 'Cost of New Construction' (CNC) index - reproduced in chapter 9, section 9.10 - until 1980 but superseded at an earlier date by the 'output price indices' referred to above (see chapter 4).

Repricing Aggregate Factor Costs. Briefly, this method consists of taking aggregate information about the cost of factor inputs (of the kind obtained in the annual census of production) and deflating it to base-year price levels through the use of appropriate factor cost indices. Comparison of the sum of factor costs at current and base-period prices yields an index of the change in output prices. The method was used by one of the authors to devise an index for Northern Ireland (Fleming, 1965) but it has not been used in Great Britain and no further discussion is offered here.

Use of Published Unit ('Measured') Rates. This method is analogous to that using unit rates extracted from priced bills of quantities, described above, the difference being that the rates used are those published in builders' price books or trade journals. These rates are meant to represent the going rates for carrying out specific items of building work. *To the extent to which this is the case* they possess certain advantages from the point of view of devising an index. As with unit rates from priced bills, they facilitate the maintenance of a comparable standard over time reflecting both changes in costs and productivity and incorporating a built-in weighting of labour and materials (they do not eliminate the comparability problem because steps still have to be taken to cope with technical change). By taking suitable combinations of rates, it is possible to devise measures of costs for different types of building and construction work. However, the problem of allowing for changes in productivity and other market conditions is not so readily overcome as the description above appears to promise. In practice, published rates are built up on the basis of labour and materials 'constants' (i.e. the quantities of materials, the estimated man-hours and possibly plant-hours required to carry out the item of work in question). Changes in productivity may be allowed for when the rates, or more particularly the constants, are revised but in practice this is likely to be infrequent. Overheads and profits are commonly allowed for in these rates through the addition of a fixed percentage. Consequently, these rates are insensitive to changes in market conditions and are more likely to reflect changes in contractors' basic factor costs rather than tender prices. The Building Research Station maintained an index of this kind until 1969 - see Chapter 9 (section 9.9).

Part B

Currently Compiled Construction Indices

3

Introduction to Current Construction Indices

Details are given in this part of the book of all the currently-compiled indices of construction costs and prices, together with a complete series of index numbers. All of the indices available are classified according to the three main types of index:

1. 'Output price' indices - considered in Chapter 4.
2. 'Tender price' indices - considered in Chapter 5.
3. 'Building cost' indices - considered in Chapter 6.

The distinction between these types of index has already been discussed. In brief, output price indices are derived from tender price indices and are used as deflators to convert official statistics of the value of contractors' output of new construction work from current prices to constant prices. Tender prices represent the cost a client must pay for a building and reflect contractors' views about the future course of costs for labour and materials during the construction period, as well as the influence of the 'tendering climate' or 'market conditions' on profit margins. 'Building costs' are the costs actually incurred by the builder in the course of his business but exclude profits and also reflect *current* costs for labour, materials etc., as opposed to the *future* costs which are incorporated in tenders. This means that, for example, in times of boom tender prices may increase at a greater rate than building costs, whilst in a depression the opposite may apply. In exceptional circumstances tender prices may actually go down whilst building costs are rising.

Tender price indices are generally compiled by comparing the prices of a proportion of the items within a number of accepted tenders during a given period of time against the price of similar items in a base schedule of rates. Each tender is indexed and the mean (geometric or arithmetic) or median of the sample becomes the index for that period. It is generally accepted that nationally there is, on average, a range of approximately 30% about the mean.

Indices for the factors used in the building cost indices are produced by official bodies. Indices for the various types of material are prepared by the Department of Trade and Industry. Indices of wages are compiled by the Department of Employment. These are given by the Department of the Environment in its *Housing and Construction Statistics* (HMSO).

Several other indices are produced to assess the recovery of increased costs. The Property Services Agency (PSA) produces indices for the various work categories used within the PSA price adjustment formula for construction contracts together with several special indices (for example the cost of providing, operating and maintaining constructional plant and equipment). Specialist bodies such as the National Association of Lift Manufacturers and others also publish indices for this purpose. These are outside the scope of this book.

The choice of index depends on the information required but it will often be found that a tender price index is more relevant to the problems faced by quantity surveyors. Amongst the advantages of a tender price index are:

It indicates the movement of the cost to the client rather than to the contractor.

It is not based on other indices, as a building cost index usually is, and therefore any inherent inaccuracies are not compounded.

It provides a measure of the level of contractors' prices over a period of time that is generally accepted by statisticians.

It gives an indication of the tendering climate at the date of tender and therefore takes into consideration not only variations in factor costs but also the effect of current economic considerations.

The index for an individual scheme will indicate its price level against the norm (as indicated by a general index of tender prices) and therefore indicate its keenness.

It indicates the effectiveness of cost planning; for instance, if the scheme has a low index compared with the norm, and its price is well above the cost limit, then it has either been badly cost-planned or the cost limit was inadequate or both.

The individual tender-based index for a scheme can be used to evaluate specific price determinants, such as location, building type, method of construction, size of contract or length of contract.

Cost planning can be improved by bringing the cost of known schemes and other historical data to a common level.

It can be used to set realistic cost limits.

There is a need for indices within the construction industry that show the movement in price of different types of buildings in different locations. If sufficient data on tender prices can be collected it may be possible to construct such indices. It is important therefore that as many building tenders as possible are indexed by the bodies that compile construction cost and price indices.

4

Output Price Indices

DOE CONSTRUCTION OUTPUT PRICE INDICES

Type of Index

Output price index (OPI).

Series: Coverage and Breakdowns

Analyses of contractors' output of new building in the public and private sectors with breakdowns by building type and by sector as follows:

Series	*Table Reference*
DOE construction output price index:	
public housing, 1970 to date	4.1
private housing, 1970 to date	4.2
public works, 1970 to date	4.3
private industrial, 1970 to date	4.4
private commercial, 1970 to date	4.5
all new construction, 1970 to date	4.6

Base Dates and Period Covered

1970	from 1970 to 1978
1975	from 1970 to 1979
1980	from 1975 to 1988
1985	from 1984 to date

The pre-1985 series are shown with 1985 = 100 in the tables below (converted arithmetically by the authors).

Frequency

Quarterly.

42   *Spon's handbook of construction cost and price indices*

Geographical Coverage

Great Britain

Type and Source of Data

PSA QSSD Tender price indices, building cost indices for materials and labour and government statistics on contractors' new orders.

Method of Compilation

The following is an extract from *Housing and Construction Statistics 1978-1988.*

'The construction output price indices are derived from tender price indices and are used as "deflators" to convert contractors' output of new construction work from current prices to constant prices. Repair and maintenance output and all direct labour output are deflated using indices based on costs of material and labour. Output of new construction work in a quarter is made up of work done on contracts let during or before that quarter: the deflator can be constructed from the value and volume of orders placed in previous quarters once adjustments have been made to tender prices for changes in material and labour costs for which reimbursement is allowed under "variation of price" clauses. Let the quarter for which the output is considered be T, and let t be any calendar quarter before that quarter. Let the value at current prices of orders placed in quarter t be X_t. Let the proportion of orders containing "variation of price" clauses be p_t and, of these contracts, let the proportions adjustable for changes in costs be M_t for materials and L_t for labour. The proportion of fixed price orders is then $(1 - p_t)$. The value of output at current prices in quarter T, Y_T, is made up of F_T from fixed price contracts and V_T from contracts adjustable for cost changes. Then

$$F_T = \sum_{t=-\infty}^{T} \alpha_{tT} X_t (1 - p_t)$$

where α_{tT} is the proportion of work done on contracts placed in quarter t being performed in quarter T. If the indices of materials costs and labour costs are I_t^M and I_t^L respectively, then

53
$$V_T = \sum_{t=-\infty}^{T} \alpha_{tT} X_t p_t \left[1 = M_t \frac{(I_T^M - I_t^M)}{I_t^M} + L_t \frac{(I_T^L - I_t^L)}{I_t^L} \right]$$

Thus the total current price output $Y_T = F_T + V_T$

$$Y_T = \sum_{t=-\infty}^{T} \alpha_{tT} X_t \left[1 + p_t \left\{ M_t \frac{(I_T^M - I_t^M)}{I_t^M} + L_t \frac{(I_T^L - I_t^L)}{I_t^L} \right\} \right]$$

If the appropriate tender price index is A_t, the volume Z_t of orders in quarter t is X_t/A_t. Thus the volume of output in quarter T, H_T, is given by

$$H_T = \sum_{t=-\infty}^{T} \alpha_{tT} Z_t = \sum_{t=-\infty}^{T} \alpha_{tT} \frac{X_t}{A_t}$$

Hence the appropriate deflator for output in a quarter T is given by:

$$D_T = \frac{Y_T}{H_T}$$

Separate output price indices are calculated for each of the five new work sectors, public housing, private housing, public non-housing, private industrial and private commercial. Within each sector contracts are further sub-divided into short, medium and long contracts, depending upon the expected duration, and the values of α_t and p_t are estimated at this lower level.

A full description of the output price indices appeared in *Economic Trends* No. 297 (July 1978). These indices superseded the Cost of New Construction (CNC) Index, (see Chapter 9, section 9.10).

Commentary

Tables 4.1 to 4.6 which follow are from 1970 with a base date of 1985 = 100, and, as stated in the extract, should be used for converting contractors' output of new construction work from current prices to constant prices.

Publications

(a) Data Source

Housing and Construction Statistics, HMSO. Quarterly and annual volumes.
Business Monitor MM17, Price Index Numbers for Current Cost Accounting,
    HMSO. Monthly.

(b) Description of Methodology

Housing and Construction Statistics, annual volume, HMSO, London and
    Economic Trends, No. 297 (July 1978), HMSO, London.

Table 4.1 DOE Construction Output Price Index: Public Housing

Base: 1985 = 100*

Year	Q1	Q2	Q3	Q4	Average
1970	17	18	18	18	18
1971	19	19	20	20	20
1972	21	22	23	23	22
1973	25	27	29	32	28
1974	34	27	39	41	38
1975	43	44	47	47	45
1976	48	48	50	51	49
1977	52	52	53	53	53
1978	54	56	58	60	57
1979	62	64	70	73	67
1980	77	80	87	89	84
1981	90	91	91	91	91
1982	90	90	90	90	90
1983	90	92	94	95	93
1984	96	97	98	98	97
1985	99	99	101	101	100
1986	102	103	104	104	103
1987	105	106	108	110	107
1988	113	116	119	122	118
1989	125	129	132**	134**	130**
1990					
1991					
1992					
1993					
1994					
1995					

Source: DOE *Housing and Construction Statistics* (HMSO).

*   Indices from 1970 to 1971 inclusive converted arithmetically by the authors from base of 1975 = 100 to 1985 = 100.

** Provisional.

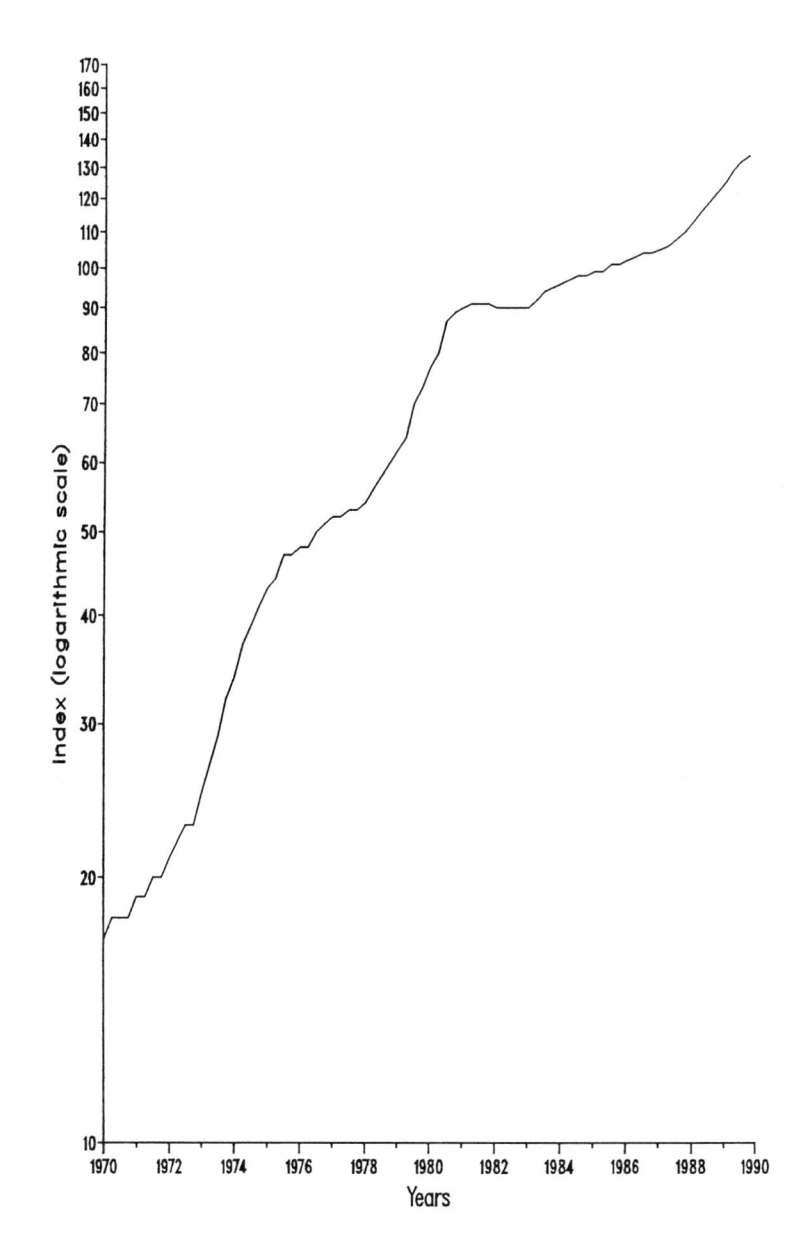

Figure 4.1 DOE Construction Output Price Index: Public Housing (1985 = 100), 1970 Q1 – 1989 Q4

Table 4.2 DOE Construction Output Price Index: Private Housing

Base: 1985 = 100*

Year	Q1	Q2	Q3	Q4	Average
1970	15	15	15	16	15
1971	16	17	17	18	17
1972	19	19	20	22	20
1973	24	26	27	29	27
1974	31	32	33	34	32
1975	36	37	38	40	38
1976	40	40	42	42	41
1977	43	44	46	47	45
1978	48	50	53	55	52
1979	57	60	65	68	63
1980	72	76	80	82	78
1981	83	84	83	82	83
1982	82	83	84	84	83
1983	86	88	91	92	89
1984	93	94	96	96	95
1985	97	99	101	103	100
1986	105	108	111	113	109
1987	116	119	123	128	122
1988	133	139	146	152	143
1989	158	163	167**	170**	165**
1990					
1991					
1992					
1993					
1994					
1995					

Source: DOE *Housing and Construction Statistics* (HMSO).

* Indices from 1970 to 1971 inclusive converted arithmetically by the authors from base of 1975 = 100 to 1985 = 100.

** Provisional.

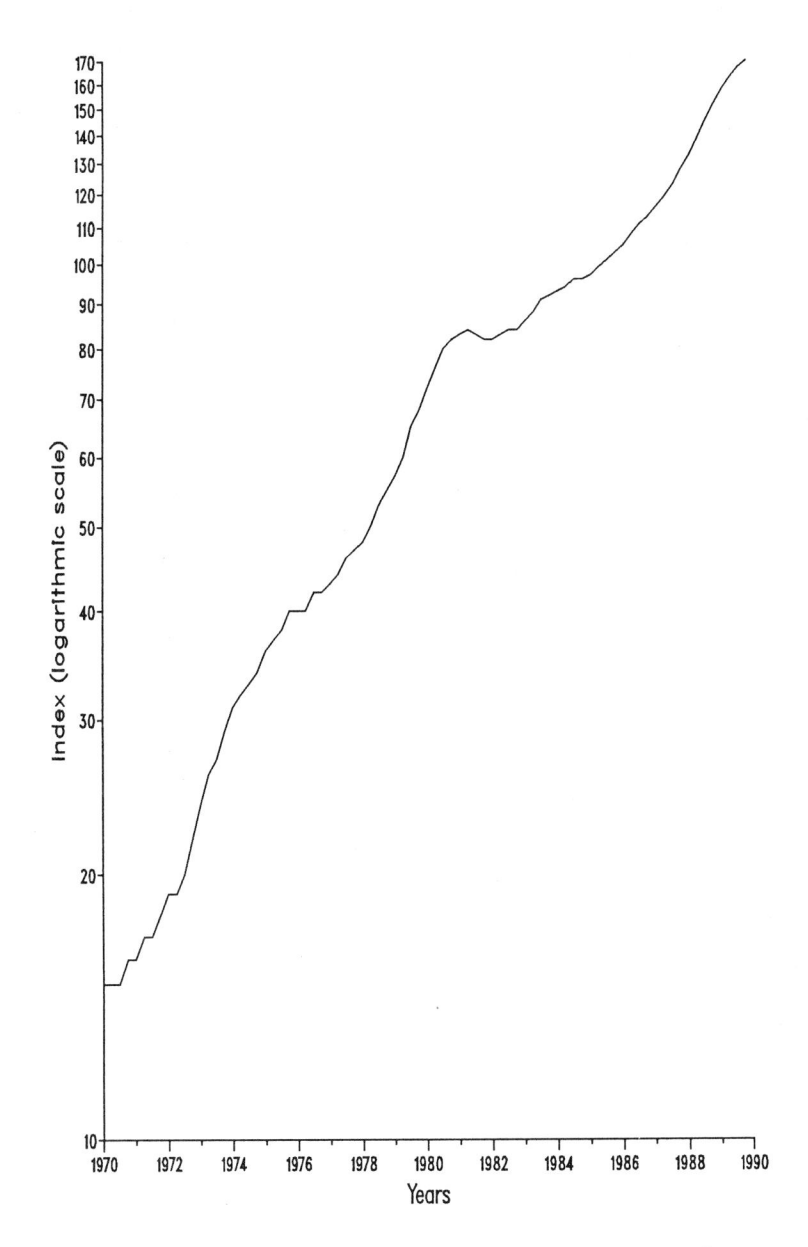

Figure 4.2 DOE Construction Output Price Index: Private Housing (1985 = 100), 1970 Q1 – 1989 Q4

Table 4.3 DOE Construction Output Price Index: Public Works

Base: 1985 = 100*

Year	Q1	Q2	Q3	Q4	Average
1970	18	19	19	19	19
1971	20	20	21	21	21
1972	22	22	23	25	23
1973	26	28	29	31	28
1974	33	35	38	40	37
1975	42	44	47	48	45
1976	49	50	51	51	50
1977	51	51	53	54	53
1978	55	57	60	62	59
1979	65	69	74	78	72
1980	84	89	97	99	92
1981	98	97	97	97	97
1982	96	96	96	95	96
1983	94	95	96	95	95
1984	95	95	97	97	96
1985	98	99	101	102	100
1986	103	102	102	102	102
1987	104	106	108	110	107
1988	113	116	120	124	118
1989	129	132	135**	137**	133**
1990					
1991					
1992					
1993					
1994					
1995					

Source: DOE *Housing and Construction Statistics* (HMSO).

*   Indices from 1970 to 1971 inclusive converted arithmetically by the authors from base of 1975 = 100 to 1985 = 100.

** Provisional.

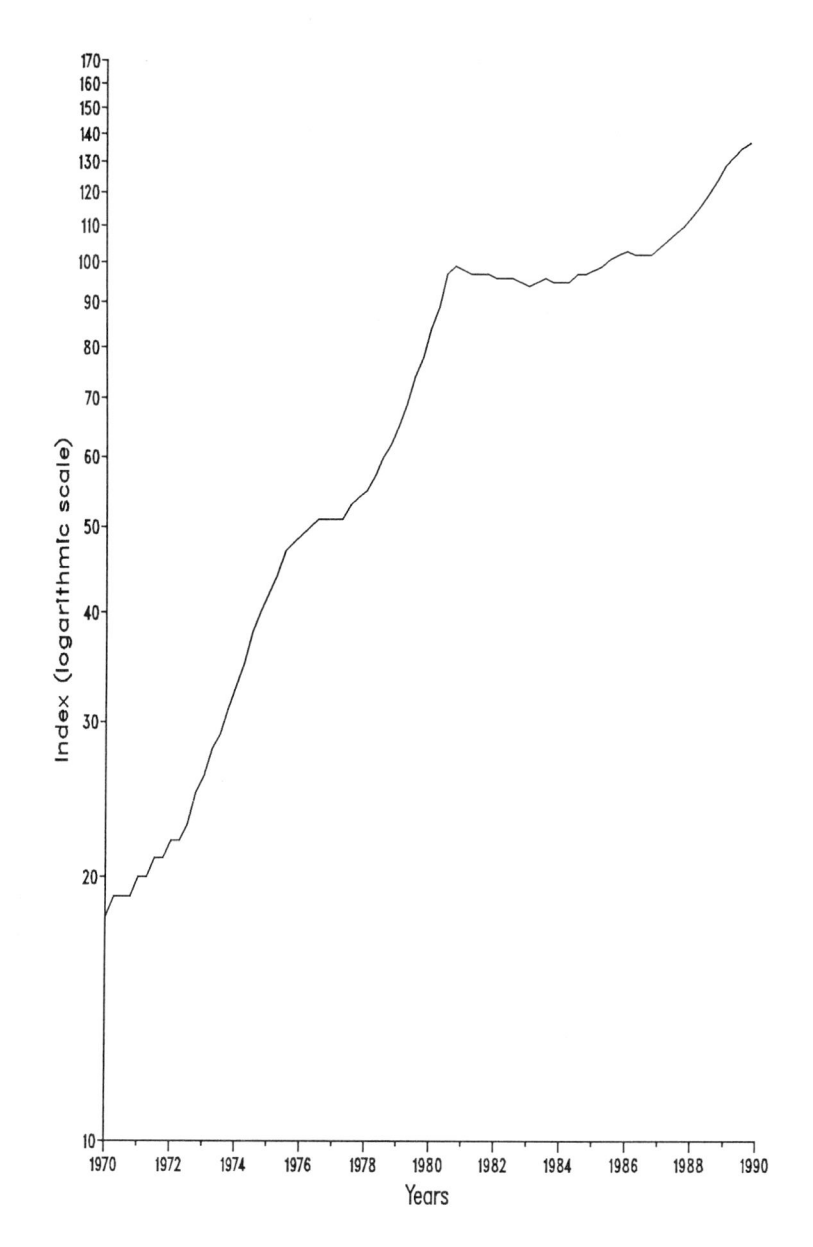

Figure 4.3 DOE Construction Output Price Index: Public Works (1985 = 100) 1970 Q1 – 1989 Q4

Table 4.4 DOE Construction Output Price Index: Private Industrial

Base: 1985 = 100*

Year	Q1	Q2	Q3	Q4	Average
1970	17	18	18	18	18
1971	19	19	20	20	20
1972	21	22	24	26	23
1973	27	29	32	35	31
1974	38	39	41	42	40
1975	43	44	45	46	44
1976	46	48	49	50	48
1977	51	51	53	55	53
1978	56	57	60	63	59
1979	66	70	78	83	74
1980	90	93	99	100	95
1981	99	98	97	95	97
1982	94	95	95	94	95
1983	93	93	94	93	93
1984	95	97	97	94	96
1985	95	100	103	102	100
1986	102	102	100	99	101
1987	101	105	108	109	106
1988	111	113	116	120	115
1989	124	127	132**	135**	130**
1990					
1991					
1992					
1993					
1994					
1995					

Source: DOE *Housing and Construction Statistics* (HMSO).

*   Indices from 1970 to 1971 inclusive converted arithmetically by the authors from base of 1975 = 100 to 1985 = 100.

**   Provisional.

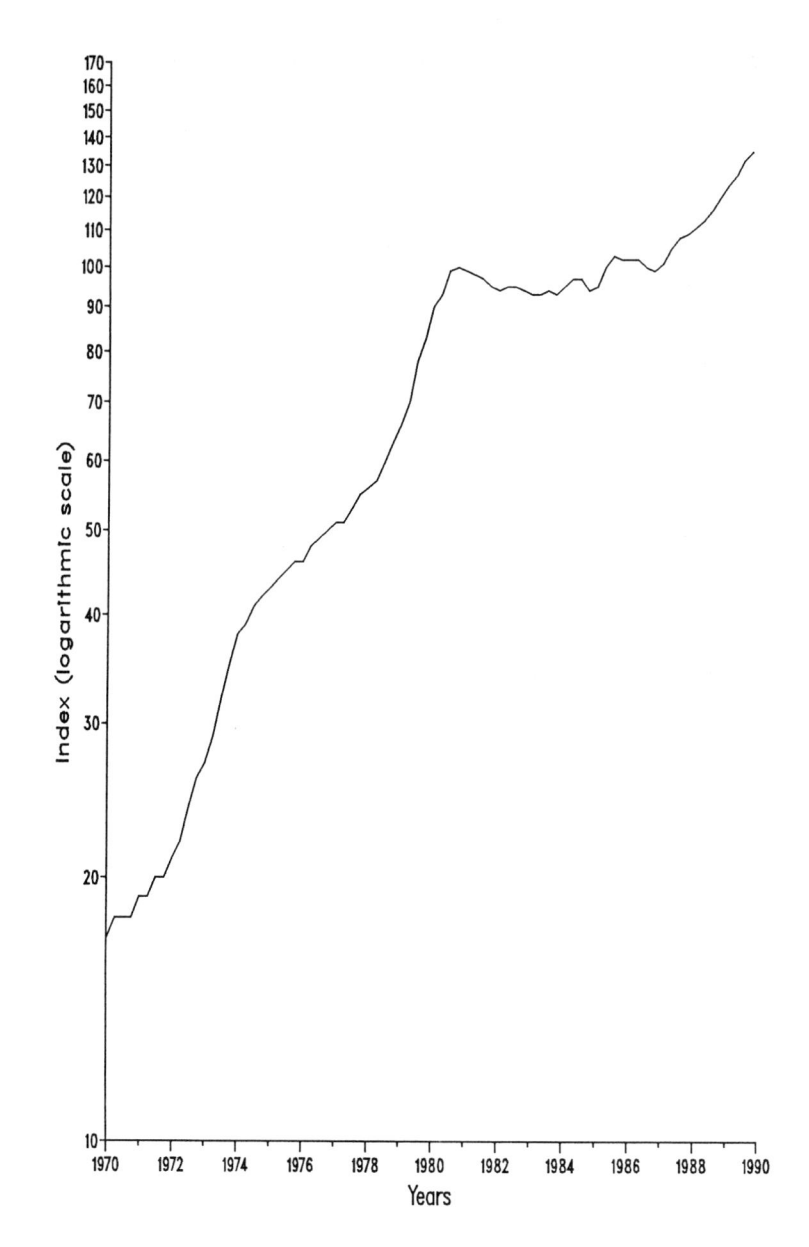

Figure 4.4 DOE Construction Output Price Index: Private Industrial (1985 = 100), 1970 Q1 – 1989 Q4

Table 4.5 DOE Construction Output Price Index: Private Commercial

Base: 1985 = 100*

Year	Q1	Q2	Q3	Q4	Average
1970	18	18	18	19	18
1971	19	20	21	21	20
1972	22	23	24	26	24
1973	27	29	32	34	30
1974	38	40	42	44	41
1975	45	46	48	48	47
1976	49	49	51	52	50
1977	53	54	55	57	55
1978	57	59	62	65	61
1979	69	73	79	85	76
1980	90	93	99	100	96
1981	100	99	97	96	98
1982	95	95	96	95	95
1983	94	94	94	93	94
1984	93	93	94	95	94
1985	96	98	102	104	100
1986	104	105	107	108	106
1987	109	109	110	113	110
1988	116	120	124	128	122
1989	131	135	137**	138**	135**
1990					
1991					
1992					
1993					
1994					
1995					

Source: DOE *Housing and Construction Statistics* (HMSO).

*   Indices from 1970 to 1971 inclusive converted arithmetically by the authors from base of 1975 = 100 to 1985 = 100.

**   Provisional.

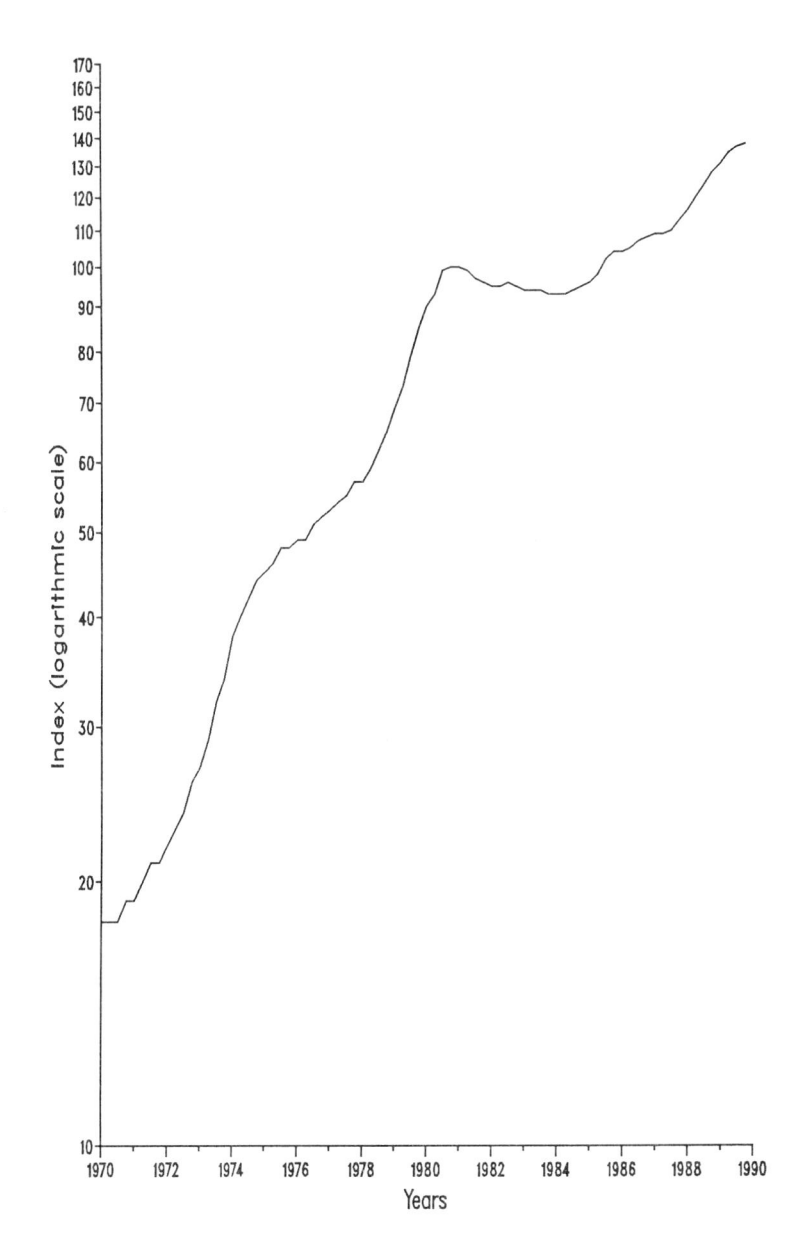

Figure 4.5 DOE Construction Output Price Index: Private Commercial (1985 = 100), 1970 Q1 – 1989 Q4

Table 4.6 DOE Construction Output Price Index: All New Construction

Base: 1985 = 100*

Year	Q1	Q2	Q3	Q4	Average
1970	17	18	18	18	18
1971	19	19	20	20	19
1972	21	21	23	25	23
1973	26	27	30	33	29
1974	35	36	38	41	37
1975	42	43	45	46	44
1976	46	47	49	50	48
1977	50	51	52	53	51
1978	54	56	59	61	57
1979	64	67	74	77	71
1980	82	86	92	94	89
1981	94	94	93	92	93
1982	91	91	92	92	92
1983	92	92	93	93	93
1984	94	95	96	96	95
1985	97	99	101	103	100
1986	104	104	105	106	105
1987	108	110	113	115	112
1988	119	122	127	132	125
1989	136	140	143**	145**	141**
1990					
1991					
1992					
1993					
1994					
1995					

Source: DOE *Housing and Construction Statistics* (HMSO).

* Indices from 1970 to 1971 inclusive converted arithmetically by the authors from base of 1975 = 100 to 1985 = 100.

** Provisional.

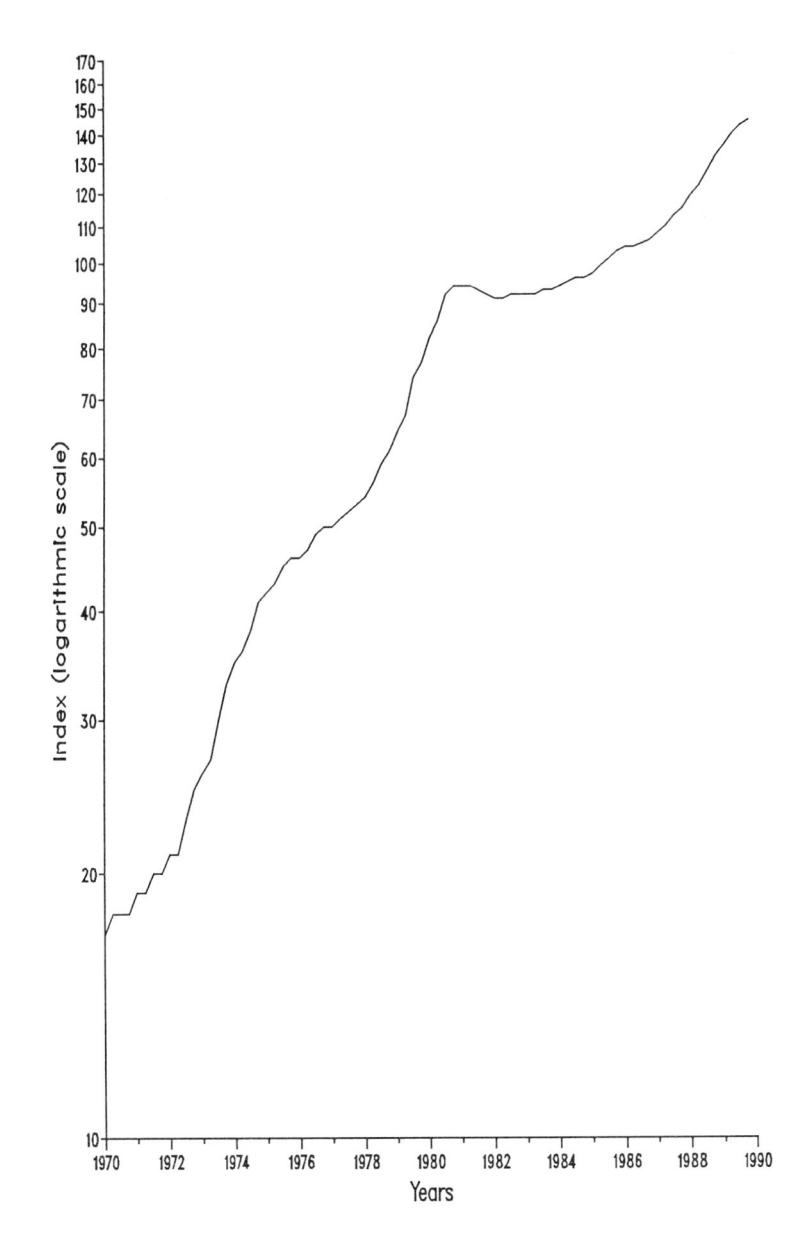

Figure 4.6 DOE Construction Output Price Index: All New Construction (1985 = 100), 1970 Q1 – 1989 Q4

5

Tender Price Indices

5.1 DOE PUBLIC SECTOR BUILDING TENDER PRICE INDICES

Type of Index

Tender price index (TPI)

Series: Coverage and Breakdowns

Analyses of new building in the public sector with breakdowns by type of contract as follows:

Series	*Table Reference*
All-in TPI, 1975 to date	5.1.1
All-in TPI, 1975 to date (value weighted)	5.1.1A
Firm Price TPI, 1975 to date	5.1.2
Firm Price TPI, 1975 to date (value weighted)	5.1.2A
Variation of Price TPI, 1975 to date	5.1.3
Variation of Price TPI, 1975 to date (value weighted)	5.1.3A

Base Dates and Period Covered

1975	from 1975 to 1979
1980	from 1975 to 1988
1985	from 1980 to date

The pre-1985 series are shown with 1985 = 100 in the tables below (converted arithmetically by either the DOE or the authors).

Frequency

Quarterly

Geographical Coverage

Great Britain

Type and Source of Data

Sample of selected tenders for new building work (excluding housing, work of a mainly civil engineering nature, mechanical and electrical work and complex works of alterations and extensions) in the public sector priced in competition, for schools, hospitals, prisons, courts, military buildings, social services buildings, and police buildings, together with contracts placed by the Property Services Agency. Details of the tenders - priced bills of quantities - are supplied from within the various Departments.

Method of Compilation

The following is based on an extract from *Housing and Construction Statistics 1978-1988*.

This index, prepared by the Quantity Surveying Services Division of the Property Services Agency, Specialist Services, measures the movement of prices in competitive tenders for building contracts in the public sector. It does not include contracts for housing work, work of a mainly civil engineering nature, mechanical and electrical work nor alterations and extensions. The principal contributors to the sample from which the index is produced are the Property Services Agency, Scottish Development Department, Department of Health, Department of Social Security, Home Office, and Department of Education and Science. Examples of building work covered in this index include schools, hospitals, prisons, courts, police stations, government offices, military buildings etc.

Each project index is prepared from price levels established by comparing prices of items to a minimum value of 25 per cent for each trade or section of the Bill of Quantities with standard base prices; information is combined by applying weights representing the proportion of the total value of the Bills of Quantities represented by each trade or section. Preliminaries and other general charges are spread proportionately over each item of the Bill of Quantities.

A quarterly value-weighted index is calculated from the price level of each project index using the full tender value as weight. Sub-indices for contracts with and without a variation of price clause are produced in a similar manner.

The index is 1980 weighted, and scaled to a value of 100 for 1985. It should be noted that this index is value-weighted and primarily intended as a deflator for construction output. An unweighted version of this index, intended for Quantity Surveyors' use, is published in the *Quantity Surveyors Information Notes*, available from Publications Sales, Building Research Establishment, Garston, Watford, WD2 7JR.

Commentary

These indices measure the change in tender price levels between one time period and another. Their main use, therefore, is in updating tenders or producing an estimate of current tender prices on the basis of historic cost information. As the published indices represent general trends, their application in any particular use needs to take account of regional differences in price levels (the PSA produce a guide to the influence of locational factors and building function factors from time to time as a study), and possibly size of contract and any specific or unusual features.

The variation of price, unweighted version, of this index (Table 5.1.3), is now used by the Department of Health to update its cost allowances. In using this index the 1975 = 100 figures are used and the index is referred to by them as 'The MIPS Index (Median Index of Public Sector Building Tender Prices)' within their organisation.

Publications

(a) Data Source

DOE, *Housing and Construction Statistics* (HMSO).

PSA *Quantity Surveyors Information Notes*, prepared by the Directorate of Building and Quantity Surveying Services of the PSA, Specialist Services, available on subscription from Publication Sales, Building Research Establishment, Garston, Watford WD2 7JR.

(b) Description of Methodology

Details of the methodology are given in *Housing and Construction Statistics, 1978-1988*, HMSO, London.

Table 5.1.1 DOE All-In Public Sector Building Tender Price Index

Base: 1985 = 100*

Year	Q1	Q2	Q3	Q4	Average
1975	43	45	44	43	44
1976	45	46	48	47	46
1977	49	52	54	56	53
1978	57	61	64	63	61
1979	68	76	81	83	77
1980	86	94	89	86	89
1981	83	87	85	86	85
1982	88	84	86	87	86
1983	87	87	90	91	89
1984	94	95	93	93	94
1985	95	98	99	99	98
1986	101	101	98	104	101
1987	106	110	109	115	110
1988	121	125	129	127	126
1989	134	135	136	130**	134**
1990					
1991					
1992					
1993					
1994					
1995					

Source: PSA *Quantity Surveyors Information Notes*.

*   Value-weighted base. The indices from 1975 to 1979 inclusive were converted arithmetically by the authors from a base of 1980 = 100 to 1985 = 100. The value weighted version of this index appears in the DOE *Housing and Construction Statistics* (see Table 5.1.1A)

**   Provisional.

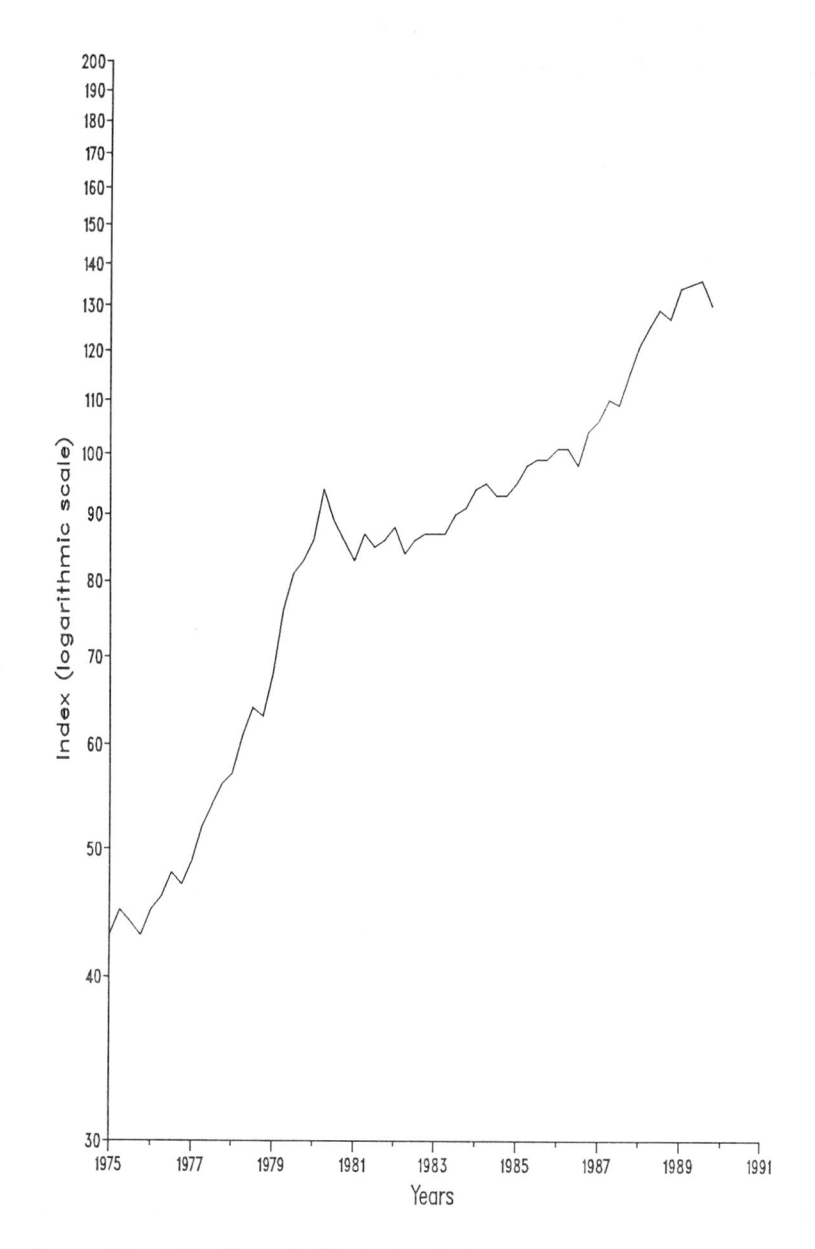

Figure 5.1.1 DOE All-in Public Sector Building Tender Price Index (1985 – 100), 1975 Q1 – 1989 Q4

Table 5.1.1A DOE All-In Public Sector Building Tender Price Index (Value Weighted Version)

Base: 1985 = 100*

Year	Q1	Q2	Q3	Q4	Average
1975	43	44	43	43	43
1976	44	45	48	47	46
1977	49	52	53	55	53
1978	56	61	63	62	61
1979	68	75	81	82	76
1980	86	94	91	90	90
1981	84	87	83	86	85
1982	90	84	87	83	86
1983	88	88	87	86	87
1984	89	92	94	90	91
1985	97	99	102	102	100
1986	102	100	100	104	102
1987	109	107	109	118	111
1988	122	125	137	134	129
1989	138	166	145	139**	147**
1990					
1991					
1992					
1993					
1994					
1995					

Source: DOE *Housing and Construction Statistics* (HMSO).

* Indices from 1975 to 1979 inclusive converted arithmetically by the authors from base of 1980 = 100 to 1985 = 100.

** Provisional.

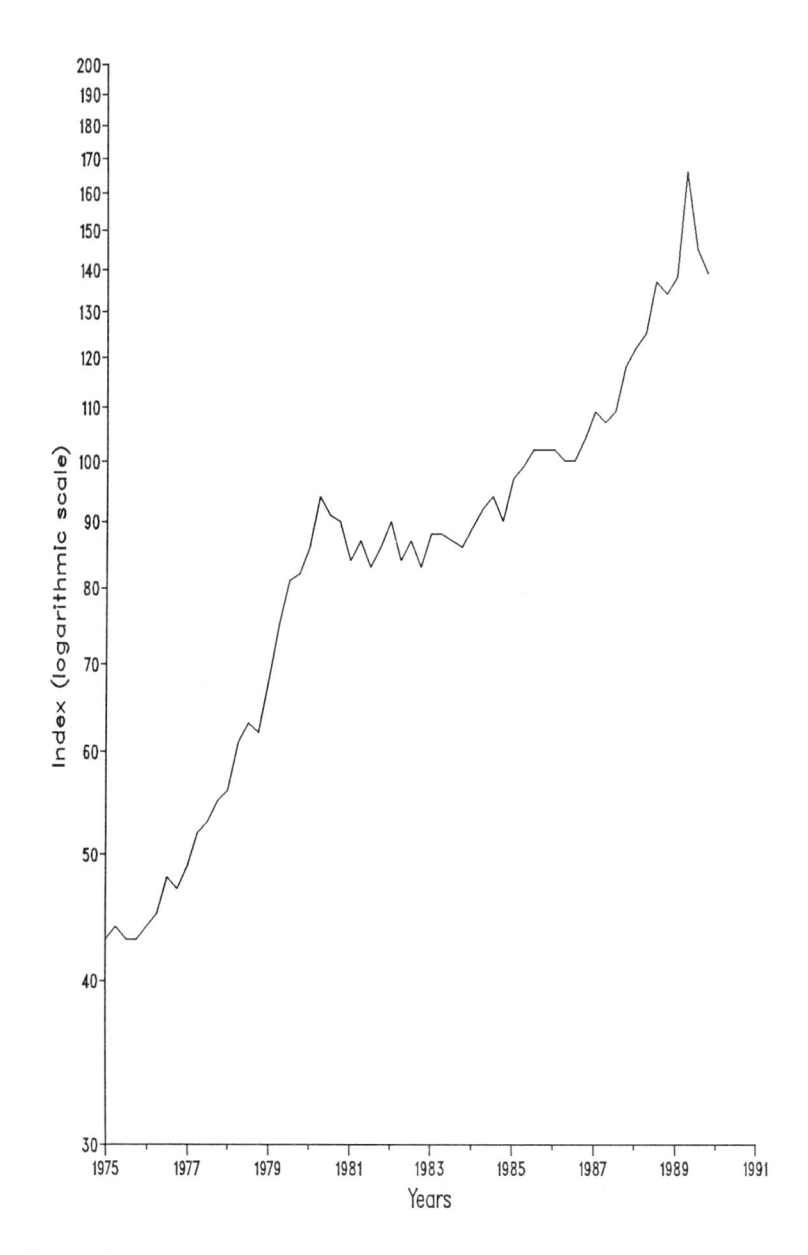

*Figure 5.1.1A DOE All-in Public Sector Building Tender Price Index –
Value Weighted Version – (1985 = 100), 1975 Q1 – 1989 Q4*

64 *Spon's handbook of construction cost and price indices*

Table 5.1.2 DOE Firm Price Public Sector Building Tender Price Index

Base: 1985 = 100*

Year	Q1	Q2	Q3	Q4	Average
1975	46	48	47	45	47
1976	47	48	50	50	49
1977	52	56	57	60	56
1978	59	61	67	72	65
1979	75	78	85	85	83
1980	96	106	95	97	99
1981	85	92	89	92	90
1982	92	91	92	90	91
1983	94	91	93	98	94
1984	103	97	97	98	99
1985	105	109	102	103	105
1986	103	102	99	105	102
1987	106	110	110	115	110
1988	122	128	131	127	127
1989	137	134	136	130**	134**
1990					
1991					
1992					
1993					
1994					
1995					

Source: PSA *Quantity Surveyors Information Notes.*

* Valued-weighted base. The indices from 1975 to 1979 inclusive were converted arithmetically by the authors from a base of 1980 = 100 to 1985 = 100. A value weighted version of this index appears in the DOE Housing and Construction Statistics (see Table 5.1.2A).

** Provisional.

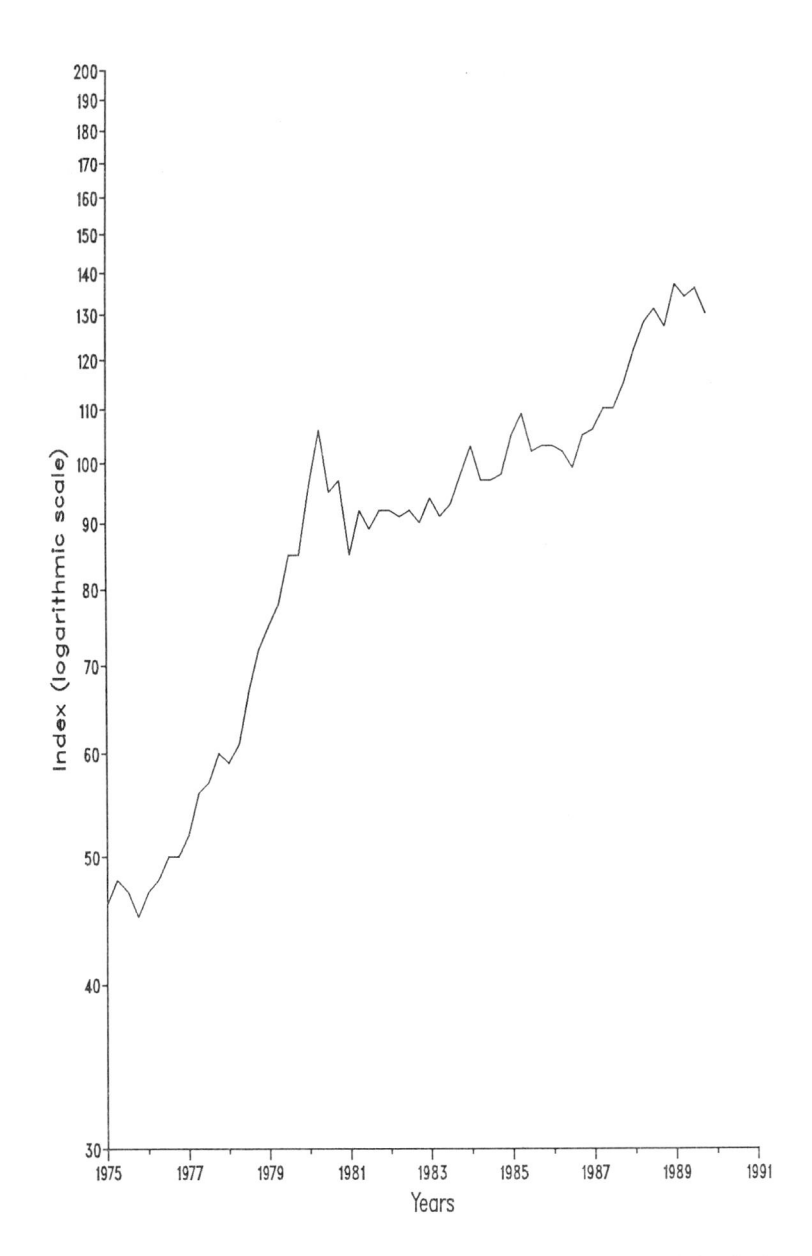

Figure 5.1.2 DOE Firm Price Public Sector Building Tender Price Index (1985 = 100), 1975 Q1 – 1989 Q4

Table 5.1.2A DOE Firm Price Public Sector Building Tender Price Index (Value Weighted Version)

Base: 1985 = 100*

Year	Q1	Q2	Q3	Q4	Average
1975	44	45	45	43	45
1976	45	45	47	47	46
1977	49	53	54	57	53
1978	56	58	63	68	62
1979	71	74	81	81	79
1980	95	100	94	93	96
1981	86	84	87	86	86
1982	89	80	89	84	86
1983	88	88	88	92	89
1984	98	91	93	91	93
1985	98	105	100	97	100
1986	93	97	95	99	96
1987	101	105	107	114	107
1988	112	125	127	126	122
1989	128	126	135	133**	131**
1990					
1991					
1992					
1993					
1994					
1995					

Source: DOE *Housing and Construction Statistics* (HMSO).

* Indices from 1975 to 1979 inclusive converted arithmetically by the authors from base of 1980 = 100 to 1985 = 100.

** Provisional.

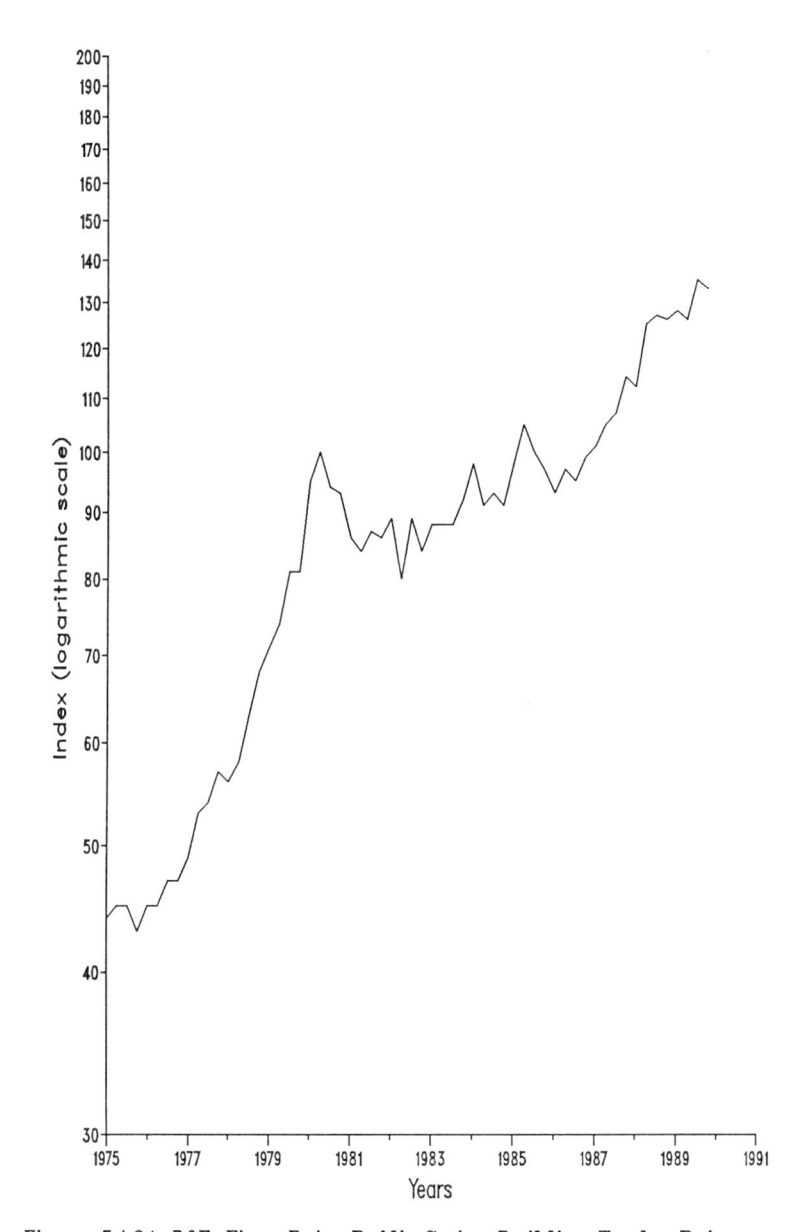

Figure 5.1.2A DOE Firm Price Public Sector Building Tender Price Index – Value Weighted Version – (1975 = 100) 1975 Q1 – 1989 Q4

Table 5.1.3 DOE Variation of Price Public Sector Building Tender Price Index

Base: 1985 = 100*

Year	Q1	Q2	Q3	Q4	Average
1975	44	45	45	45	45
1976	46	48	49	49	48
1977	49	52	52	56	52
1978	55	60	63	61	59
1979	66	64	79	79	74
1980	83	88	87	85	86
1981	83	84	82	83	83
1982	85	81	83	83	83
1983	84	85	87	89	86
1984	87	93	92	91	91
1985	92	91	96	96	94
1986	97	95	94	98	96
1987	102	106	98	117	106
1988	110	108	123	126	117
1989	116	153†	124	124**	129**
1990					
1991					
1992					
1993					
1994					
1995					

Source: PSA *Quantity Surveyors Information Notes*.

* Value-weighted base. The indices from 1975 to 1979 inclusive were converted arithmetically by the authors from a base of 1980 = 100 to 1985 = 100. A value weighted version of this index appears in the DOE Housing and Construction Statistics (see Table 5.1.3A)
** Provisional.
† Few projects predominately in the South East.

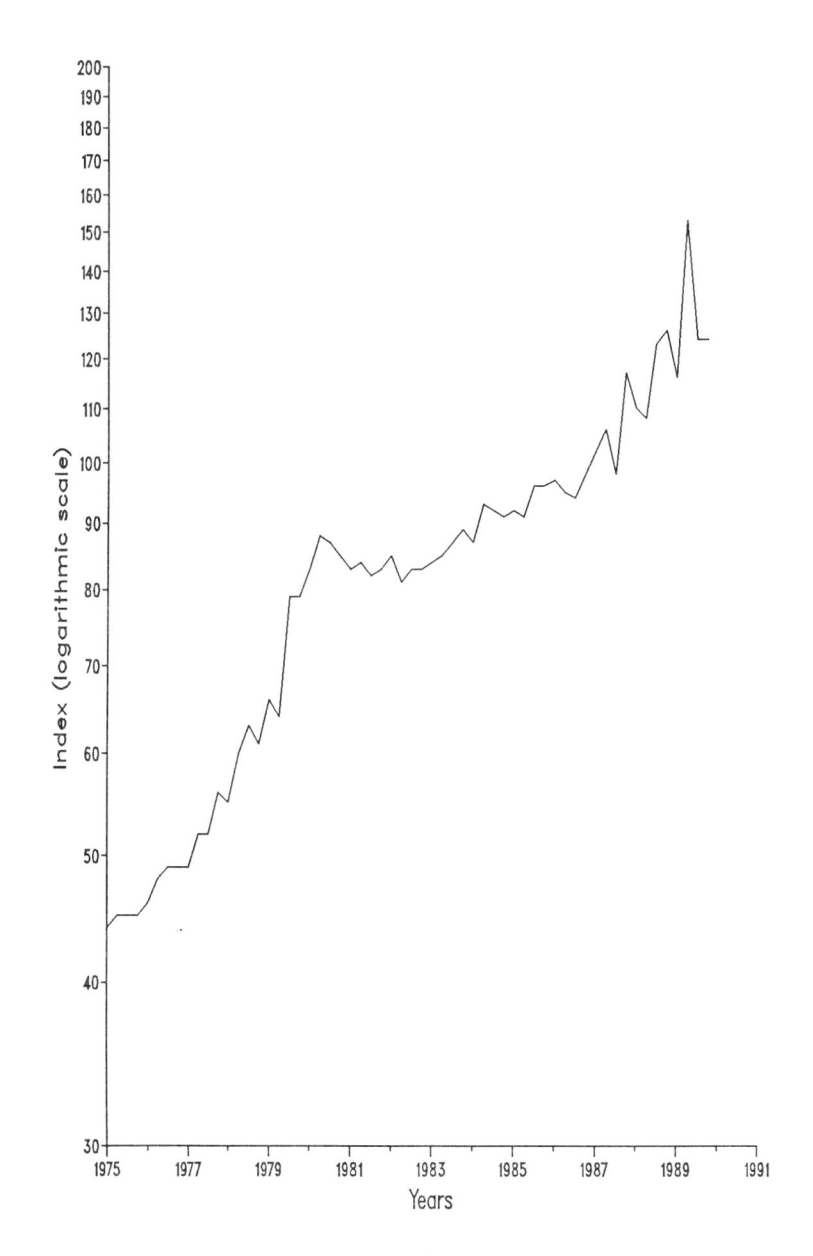

Figure 5.1.3 DOE Variation of Price Public Sector Building Tender Price Index (1985 = 100), 1975 Q1 – 1989 Q4

Table 5.1.3A DOE Variation of Price Public Sector Building Tender Price Index (Value Weighted Version)

Base: 1985 = 100*

Year	Q1	Q2	Q3	Q4	Average
1975	46	47	47	47	47
1976	48	50	52	51	50
1977	51	54	54	58	54
1978	57	63	65	64	62
1979	69	67	82	83	77
1980	88	95	94	91	92
1981	87	88	84	88	87
1982	92	86	88	85	88
1983	89	89	88	88	89
1984	89	93	96	92	93
1985	99	98	102	101	100
1986	109	100	99	104	103
1987	111	106	107	114	110
1988	126	119	140	136	130
1989	141	202	147	134**	156**
1990					
1991					
1992					
1993					
1994					
1995					

Source: DOE *Housing and Construction Statistics* (HMSO).

* Indices from 1975 to 1979 inclusive converted arithmetically by the authors from base of 1980 = 100 to 1985 = 100.

** Provisional.

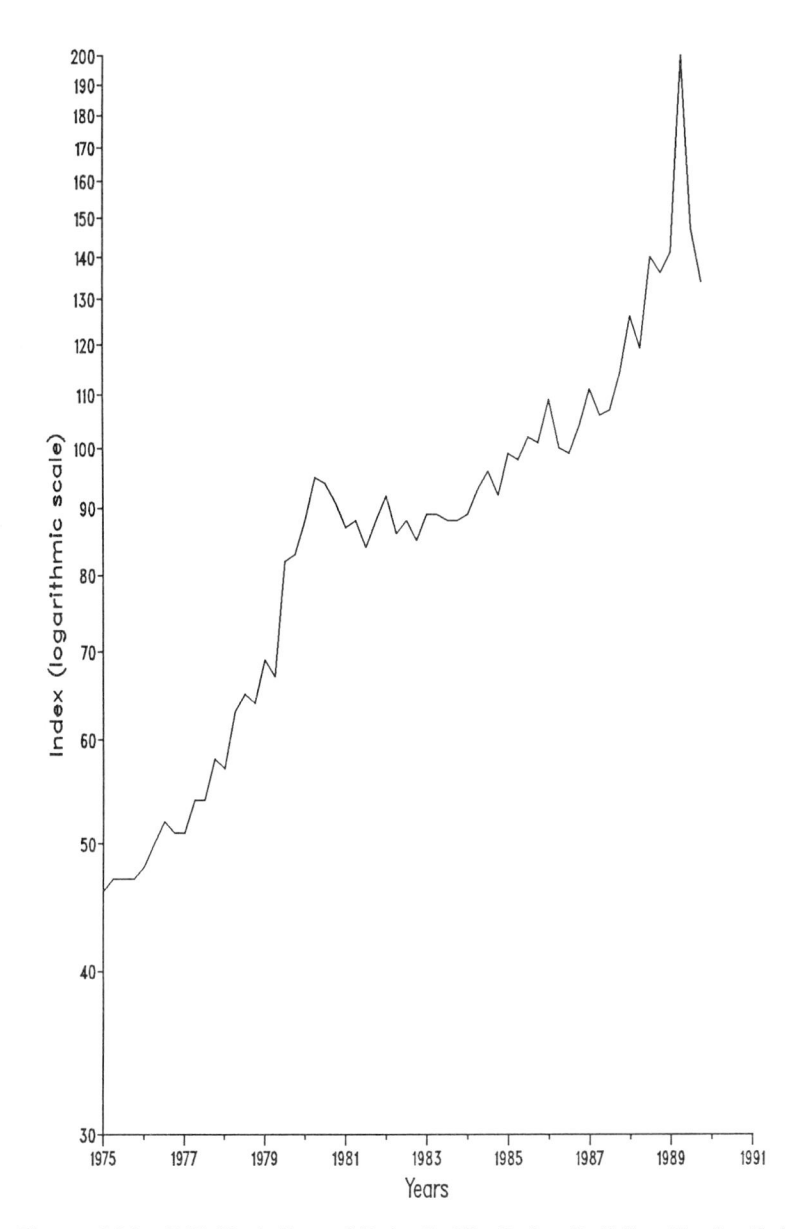

Figure 5.1.3A DOE Variation of Price Public Sector Building Tender Price Index − Value Weighted Version − (1975 = 100)
1975 Q1 − 1989 Q4

5.2 PSA QSSD TENDER PRICE INDICES

Type of Index

Tender price index (TPI)

Series: Coverage and Breakdowns

Analyses of new building within the Property Services Agency with breakdowns by type of contract as follows:

Series	*Table Reference*
All-in TPI, 1968 to date	5.2.1
Firm Price TPI, 1974 Q3 to date	5.2.2
Variation of Price TPI, 1974 Q3 to date	5.2.3

Base Dates and Period Covered

1968 Q1	from 1968 to 1979
1970	from 1970 to 1979
1975	from 1975 to 1988
1985	from 1980 to date

The pre-1985 series are shown with 1985 = 100 in the tables below (converted arithmetically by either the PSA or the authors).

Frequency

Quarterly

Geographical Coverage

Great Britain

Type and Source of Data

PSA tenders for new building work let in competition based on Bills of Quantities (excluding housing, work of a mainly civil engineering nature, mechanical and electrical work and complex works of alterations and extensions). Since about 1988, the sample for the Variation of Price TPI and the All-in TPI have been supplemented by work from other government departments.

Method of Compilation

The following is an extract from *Housing and Construction Statistics 1978-1988*, adjusted to refer specifically to this index which is similar to the DOE Public Sector Tender Price Index (see section 5.1).

'The index is prepared from price levels established for each accepted tender by comparing prices of items to a minimum value of 25 per cent for each trade or section of the Bill of Quantities with standard base prices; information is combined by applying weights representing the proportion of the total value of the Bills of Quantities represented by each trade or section. Preliminaries and other general charges are spread proportionately over each item of the Bill of Quantities. The published index numbers are the median of the project index numbers for each quarter.'

Commentary

The indices measure the change in tender price levels between one time period and another. Their main use, therefore, is in updating tenders or producing an estimate of current tender prices on the basis of historical cost information. As the published indices represent general trends, their application in any particular use needs to take account of regional differences in price levels (the PSA produce a guide to the influence of locational factors and building function factors from time to time as a study), and possibly size of contract and any specific or unusual features.

Publications

(a) Data Source

PSA *Quantity Surveyors Information Notes*, prepared by the Directorate of Building and Quantity Surveying Services of the PSA Specialist Services, available on subscription from Publication Sales, Building Research Establishment, Garston, Watford WD2 7JR.

(b) Description of Methodology

Details of the methodology are not freely available but are broadly as described above.

74 *Spon's handbook of construction cost and price indices*

Table 5.2.1 PSA QSSD All-In Index of Building Tender Prices

Base: 1985 = 100*

Year	Q1	Q2	Q3	Q4	Average
1968	17	18	18	18	18
1969	18	18	18	18	18
1970	19	20	21	21	20
1971	22	23	23	23	24
1972	26	27	29	30	28
1973	33	37	43	46	40
1974	46	47	46	45	46
1975	45	46	46	45	46
1976	47	48	50	49	49
1977	51	54	54	58	54
1978	57	60	64	68	62
1979	72	76	83	91	81
1980	91	101	93	97	96
1981	85	89	86	90	88
1982	89	88	85	84	87
1983	89	86	88	89	88
1984	90	94	93	88	91
1985	93	99	98	98	97
1986	99	101	97	100	99
1987	102	106	106	109	106
1988	108	122	127	128	121
1989	133	143†	135		
1990					
1991					
1992					
1993					
1994					
1995					

Source: PSA *Quantity Surveyors Information Notes.*

* Indices from 1968 to 1979 inclusive converted arithmetically by either the DOE or the authors from bases of 1968 Q1 = 100 or 1975 = 100 to 1985 = 100.

† Based on a very small sample.

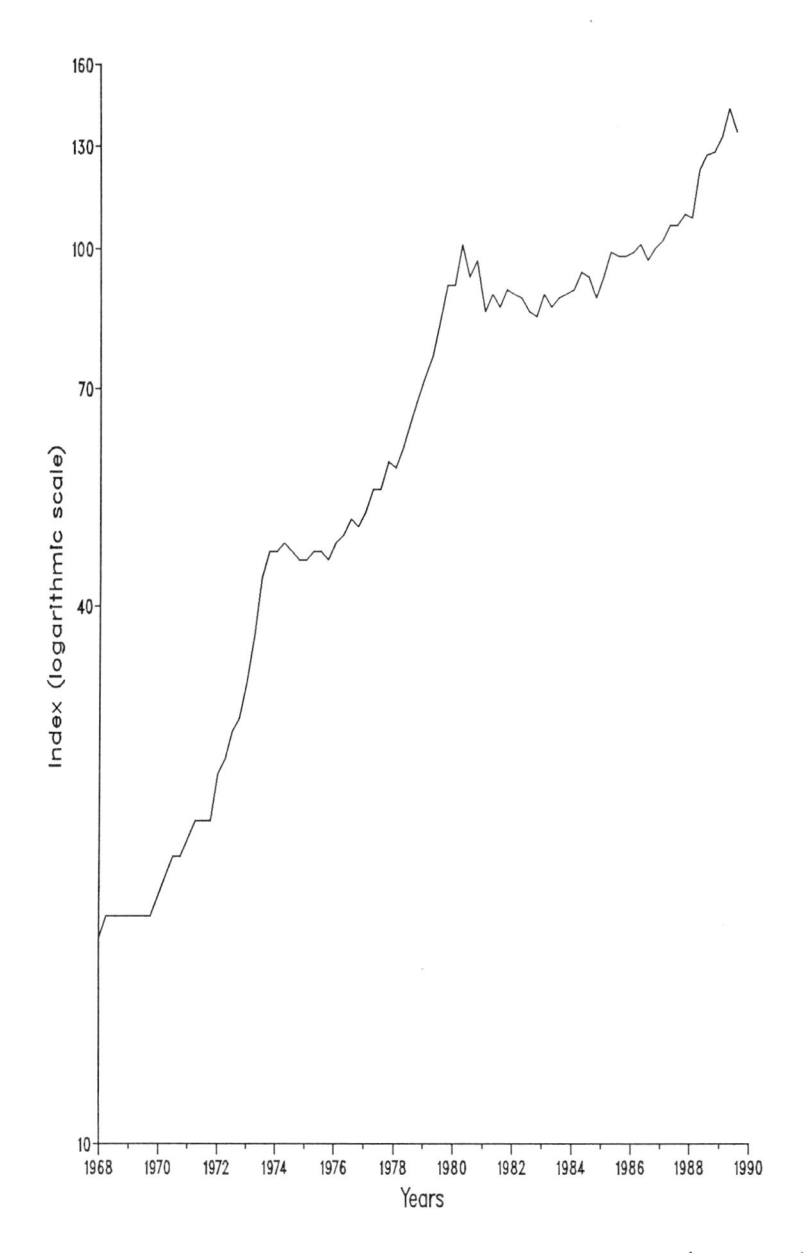

Figure 5.2.1 PSA QSSD All-in Index of Building Tender Prices (1985 = 100) 1968 Q1 – 1989 Q3

Table 5.2.2 PSA QSSD Firm Price Index of Building Tender Prices

Base: 1985 = 100*

Year	Q1	Q2	Q3	Q4	Average
1974			48	46	-
1975	46	48	47	45	47
1976	47	48	50	49	48
1977	52	55	56	60	56
1978	58	61	65	70	64
1979	75	79	84	94	83
1980	95	104	99	109	102
1981	92	97	92	99	95
1982	89	91	91	90	90
1983	99	92	93	94	95
1984	97	97	98	92	96
1985	103	107	101	102	103
1986	103	106	101	103	103
1987	103	109	112	105	107
1988	107	130	130	133	125
1989	143	137	142	146	142
1990					
1991					
1992					
1993					
1994					
1995					

Source: PSA *Quantity Surveyors Information Notes.*

* Indices from 1974 to 1979 inclusive converted arithmetically by the authors from bases of 1975 = 100 to 1985 = 100.

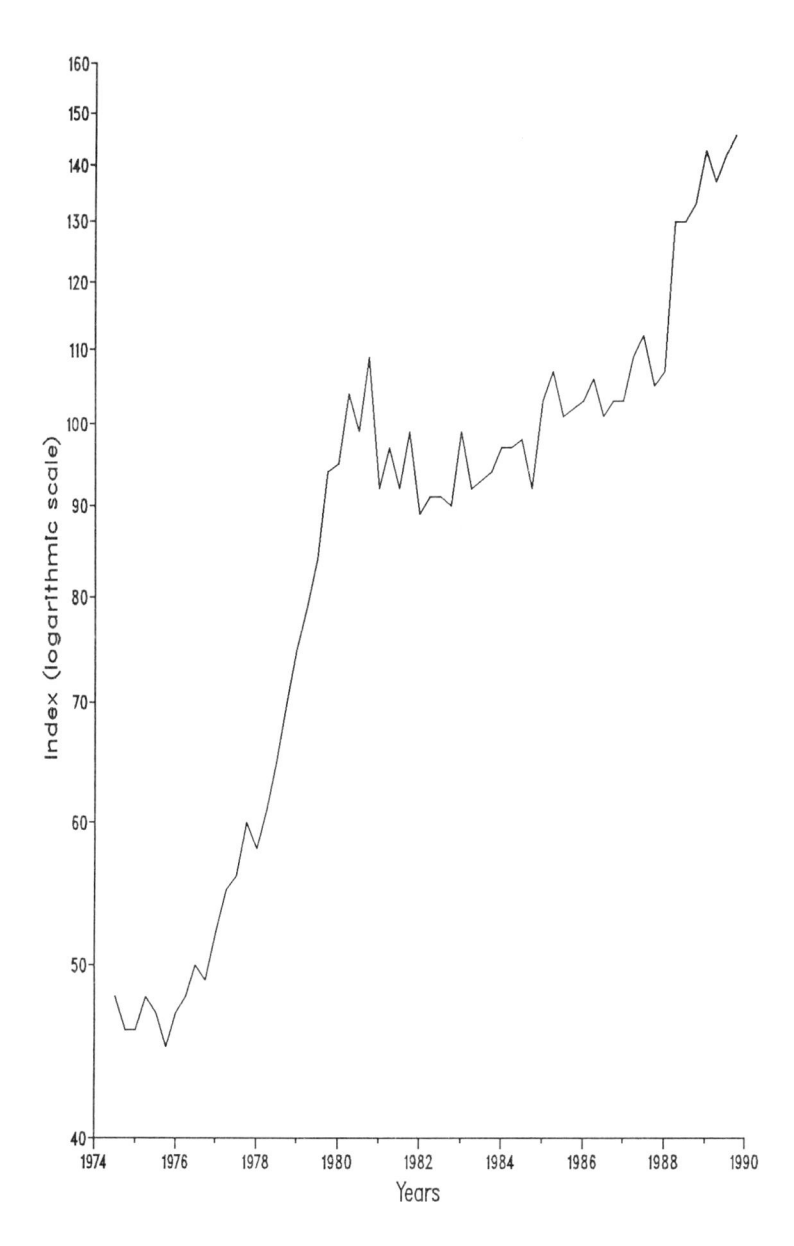

Figure 5.2.2 PSA QSSD Firm Price Index of Building Tender Prices (1985 = 100), 1974 Q3 – 1989 Q4

Table 5.2.3 PSA QSSD Variation of Price Index of Building Tender Prices

Base: 1985 = 100*

Year	Q1	Q2	Q3	Q4	Average
1974			46	44	–
1975	45	46	46	45	46
1976	47	48	50	49	49
1977	49	53	53	56	53
1978	55	58	62	65	60
1979	68	73	80	87	77
1980	84	96	84	83	87
1981	79	82	80	82	81
1982	89	86	82	81	85
1983	83	82	85	85	84
1984	84	92	88	85	87
1985	83	91	94	94	91
1986	95	93	92	96	94
1987	100	100	95	115	103
1988	109	108	123	121	115
1989	116	153†	124		
1990					
1991					
1992					
1993					
1994					
1995					

Source: PSA *Quantity Surveyors Information Notes.*

*   Indices from 1974 to 1979 inclusive converted arithmetically by the authors from bases of 1975 = 100 to 1985 = 100.

†   Based on a very small sample.

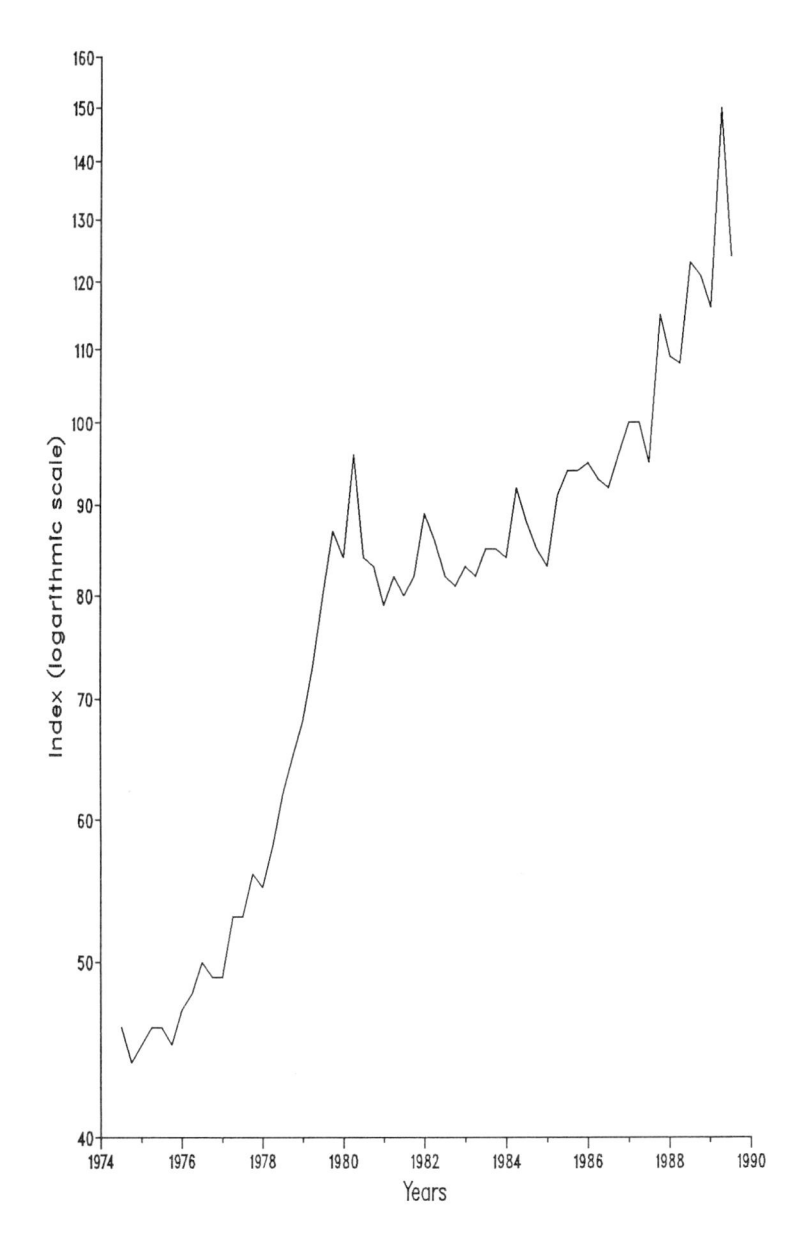

Figure 5.2.3 PSA QSSD Variation of Price Index of Building Tender Prices (1985 = 100), 1974 Q3 – 1989 Q3

5.3 BCIS TENDER PRICE INDICES

Type of Index

Tender price index (TPI).

Series: Coverage and Breakdowns

Analyses of new building in the public and private sectors with breakdowns by building type, by sector, by type of contract and by region as follows:

Series	Table Reference
All-in TPI, United Kingdom, 1974 to date	5.3.1
Firm Price TPI, 1974 to date	5.3.2
Fluctuating TPI, 1974 to date	5.3.3
South East TPI, 1984 to date	5.3.4
Rest of UK TPI, 1984 to date	5.3.5
Public Sector TPI, 1984 to date	5.3.6
Housing TPI, 1984 to date	5.3.7
Private Sector TPI, 1984 to date	5.3.8
Private Commercial TPI, 1984 to date	5.3.9
Private Industrial TPI, 1984 to date	5.3.10

Base Dates and Period Covered

1974 Q1	from 1974 to 1988
1985	from 1984 to date

The pre-1985 series are shown with 1985 = 100 in the tables below (converted arithmetically by the authors) except for series which are no longer produced.

Frequency

Quarterly

Geographical Coverage

United Kingdom.

Type and Source of Data

Sample of selected tenders for new building work (excluding housing - except for a separate housing index, given in Table 5.3.7) in the public and private sectors priced in competition or by negotiation with contract sums exceeding a defined level (currently £50,000). Details of the tenders - priced bills of quantities - are supplied by BCIS members who are requested on a random sample basis to submit the Bills of Quantities for their latest accepted tender.

Method of Compilation

The method used is that of re-pricing a sample number of items selected from each trade or section of bills of quantities, the value totalling 25 per cent of the trade or section, for current tenders on the basis of a standard schedule of rates. Individual relationships obtained from this sampling procedure are then combined to form an index for each trade or section and these trade or section indices are weighted by the trade or section totals to produce a 'project index'. There are a number of exceptions to these general rules. The Tender Price Indices are quarterly indices. Each project index is allocated to a quarter by either date of tender or base month (as applicable) of the scheme. The geometric mean of the project index figures for each quarter forms the published index figure. Separate indices, as shown in paragraph 5.3.2 above, are published.

The individual project index numbers are very variable and an attempt is made to base published indices on at least 80 projects within each quarter in order to reduce sampling error, but this is not always possible. Details of sample size are given in the primary source of the published indices (available on subscription - see below). The 1985 series is smoothed to take account of variations in the sample with regard to location, contract size and method of procurement.

Commentary

The indices measure the change in tender price levels between one time period and another. Their main use, therefore, is in updating tenders or producing an estimate of current tender prices on the basis of historic cost information. As the published indices represent general trends, their application in any particular use needs to take account of regional differences in price levels (the BCIS produce a guide to the influence of locational factors in BCIS *Quarterly Review of Building Prices*), and possibly size of contract and any specific or unusual features.

Publications

(a) Data Source

The BCIS *Quarterly Review of Building Prices* gives the All-in, Private, Public, Commercial, Industrial, South East and Rest of UK TPIs and is available on subscription from Building Cost Information Service, 86/87 Clarence Street, Kingston-upon-Thames, Surrey, KT1 1RB. The remainder are available in the full subscription service and on-line service from the Building Cost Information Service as above.

(b) Description of Methodology

Details of the methodology are given in 'Cost study F34' available from the Building Cost Information Service as above.

Table 5.3.1 BCIS All-In Tender Price Index

Base: 1985 = 100*

Year	Q1	Q2	Q3	Q4	Average
1974	41	41	41	41	41
1975	43	42	43	43	43
1976	46	45	47	48	46
1977	50	54	55	53	53
1978	56	60	63	67	62
1979	71	74	82	87	79
1980	88	92	94	89	91
1981	87	87	86	84	86
1982	89	88	86	87	88
1983	88	88	88	90	89
1984	94	96	95	96	95
1985	97	102	99	103	100
1986	101	102	103	105	103
1987	108	106	109	117	110
1988	119	123	128	128	125
1989	135	135	141**	142**	138**
1990					
1991					
1992					
1993					
1994					
1995					

Source: BCIS.

* Indices from 1974 to 1983 inclusive converted arithmetically by the authors from base of 1974 Q1 = 100 to 1985 = 100.

** Provisional.

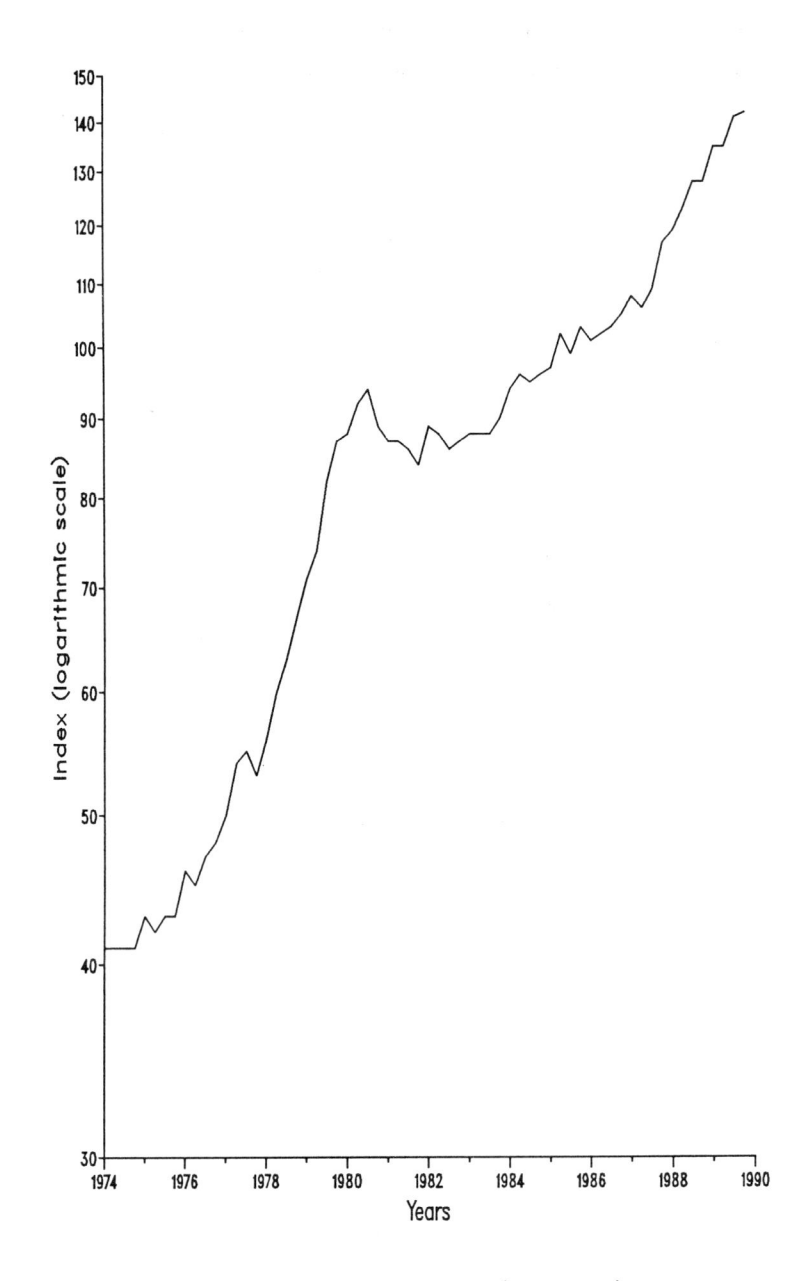

Figure 5.3.1 BCIS All-in Tender Price Index (1985 = 100), 1974 Q1 – 1989 Q4

Table 5.3.2 BCIS Firm Price Tender Price Index

Base: 1985 = 100*

Year	Q1	Q2	Q3	Q4	Average
1974	41	40	41	43	41
1975	44	44	41	43	43
1976	47	45	46	50	47
1977	49	56	55	53	53
1978	58	60	64	67	62
1979	72	75	80	89	79
1980	89	95	97	90	93
1981	88	88	84	82	86
1982	90	89	86	88	88
1983	90	88	87	92	89
1984	93	95	94	96	95
1985	97	102	99	103	100
1986	100	102	104	105	103
1987	107	105	108	116	109
1988	118	121	126	125	123
1989	134	133**	140**		
1990					
1991					
1992					
1993					
1994					
1995					

Source: BCIS.

* Indices from 1974 to 1983 inclusive converted arithmetically by the authors from base of 1974 Q1 = 100 to 1985 = 100.

** Provisional.

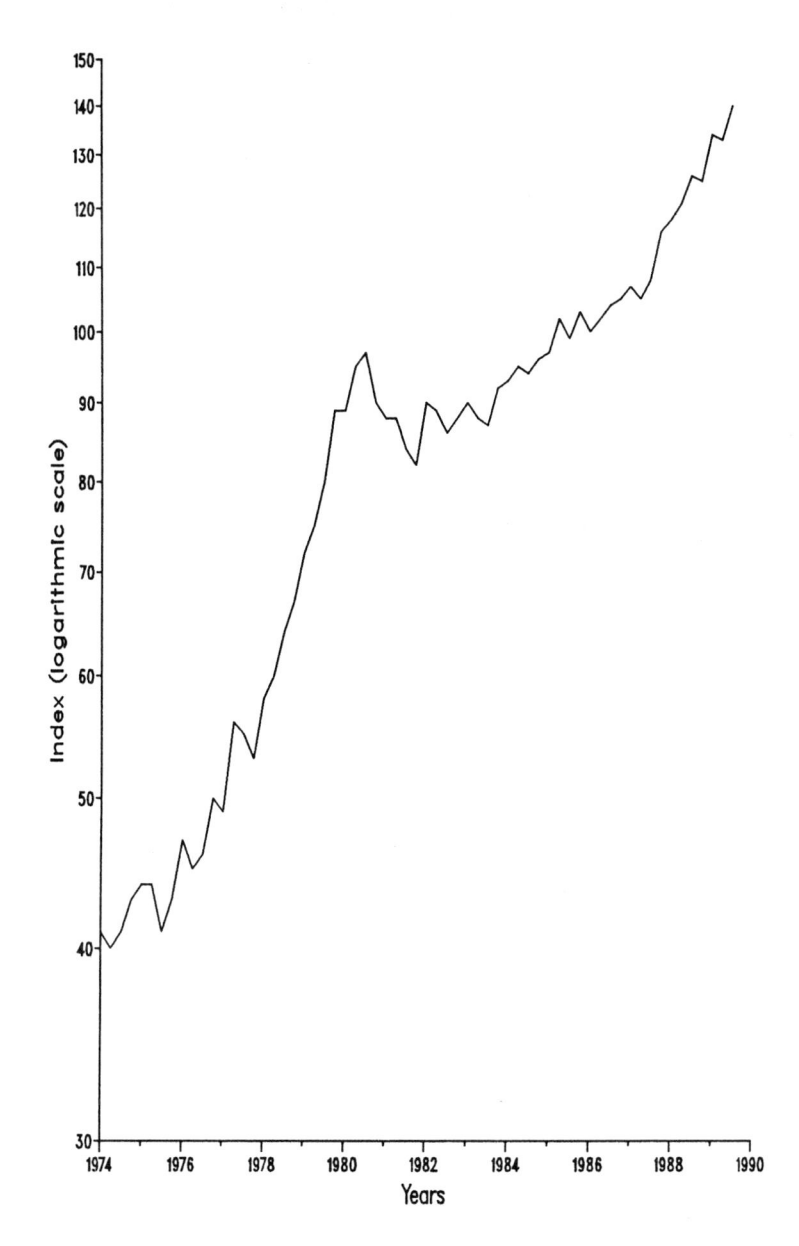

Figure 5.3.2 BCIS Firm Price Tender Price Index (1985 = 100), 1974 Q1 – 1989 Q3

Table 5.3.3 BCIS Fluctuating Tender Price Index

Base: 1985 = 100*

Year	Q1	Q2	Q3	Q4	Average
1974	43	43	43	43	43
1975	44	44	47	46	45
1976	47	46	48	49	48
1977	52	53	56	55	54
1978	58	63	65	69	64
1979	73	76	87	90	82
1980	92	93	92	92	92
1981	91	89	91	88	90
1982	89	88	87	88	88
1983	86	92	91	89	90
1984	96	98	100	98	98
1985	98	104	98	102	100
1986	103	103	97	108	103
1987	121	109	117	122	117
1988	124	135**	137**	146**	136**
1989	134**				
1990					
1991					
1992					
1993					
1994					
1995					

Source: BCIS.

* Indices from 1974 to 1983 inclusive converted arithmetically by the authors from base of 1974 Q1 = 100 to 1985 = 100.

** Provisional.

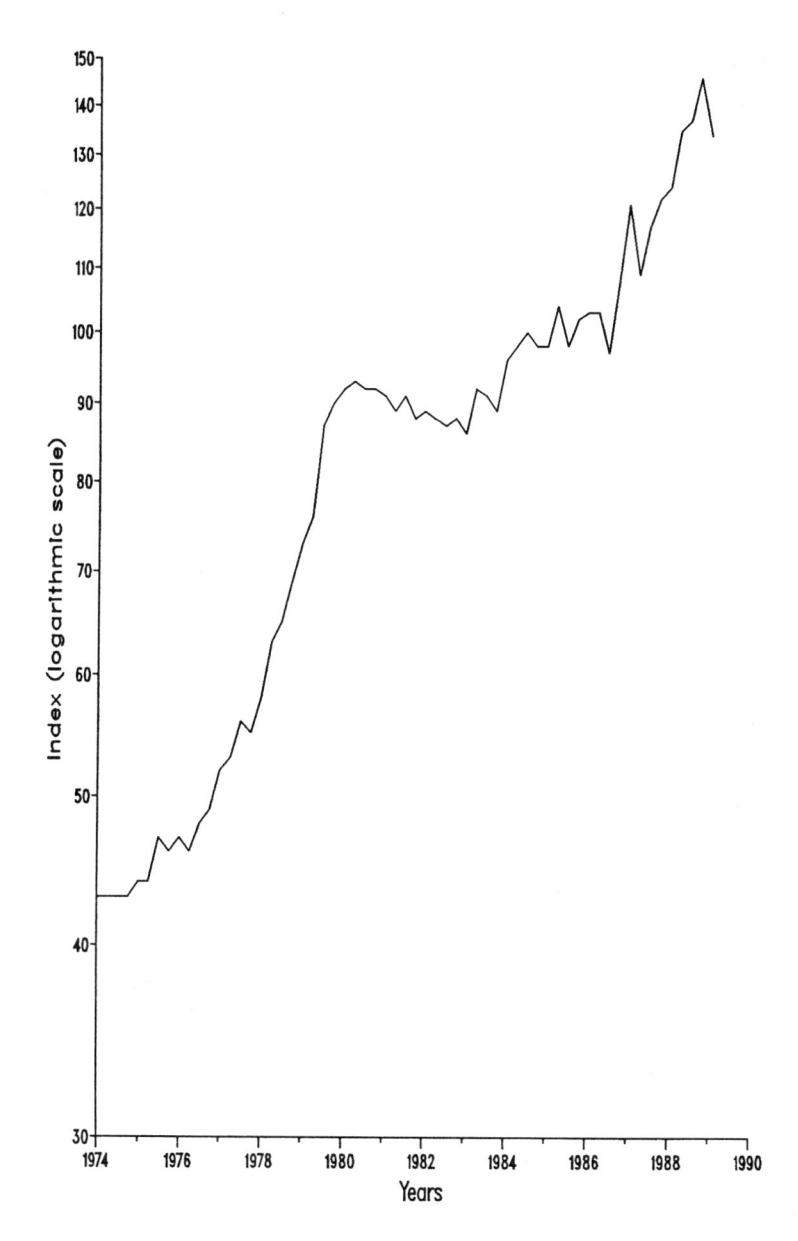

Figure 5.3.3 BCIS Fluctuating Tender Price Index (1985 = 100), 1974 Q1 – 1989 Q1

Table 5.3.4 BCIS South East Tender Price Index

Base: 1985 = 100

Year	Q1	Q2	Q3	Q4	Average
1984	89	100	93	95	94
1985	95	103	99	105	100
1986	102	105	106	111	106
1987	112	107	115	122	114
1988	123	134	137	135	132
1989	145	141**	140**		
1990					
1991					
1992					
1993					
1994					
1995					

Source: BCIS.

** Provisional.

Table 5.3.5 BCIS Rest of UK Tender Price Index

Base: 1985 = 100

Year	Q1	Q2	Q3	Q4	Average
1984	97	94	98	97	97
1985	99	102	99	101	100
1986	100	100	100	102	100
1987	105	105	106	115	108
1988	114	114	121	119	117
1989	128	130**	142**		
1990					
1991					
1992					
1993					
1994					
1995					

Source: BCIS.

** Provisional.

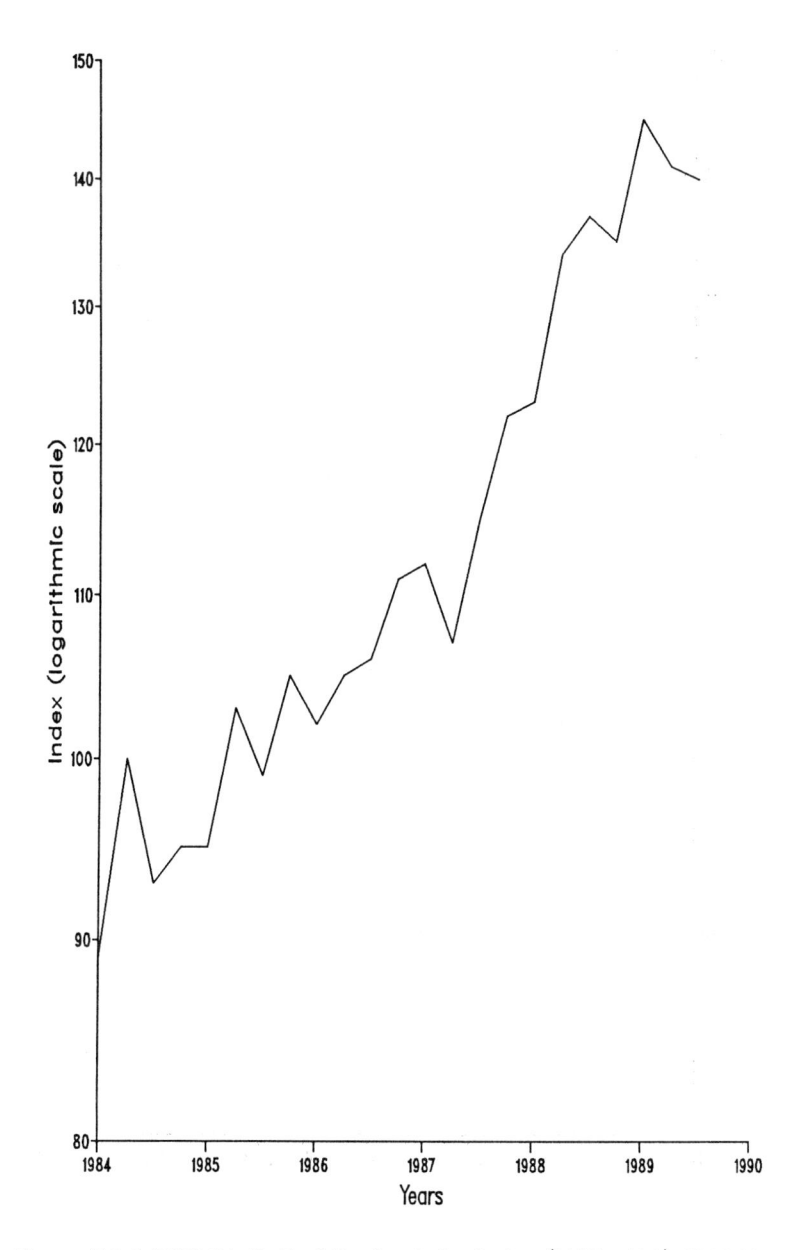

Figure 5.3.4 BCIS South East Tender Price Index (1985 = 100), 1984 Q1 – 1989 Q3

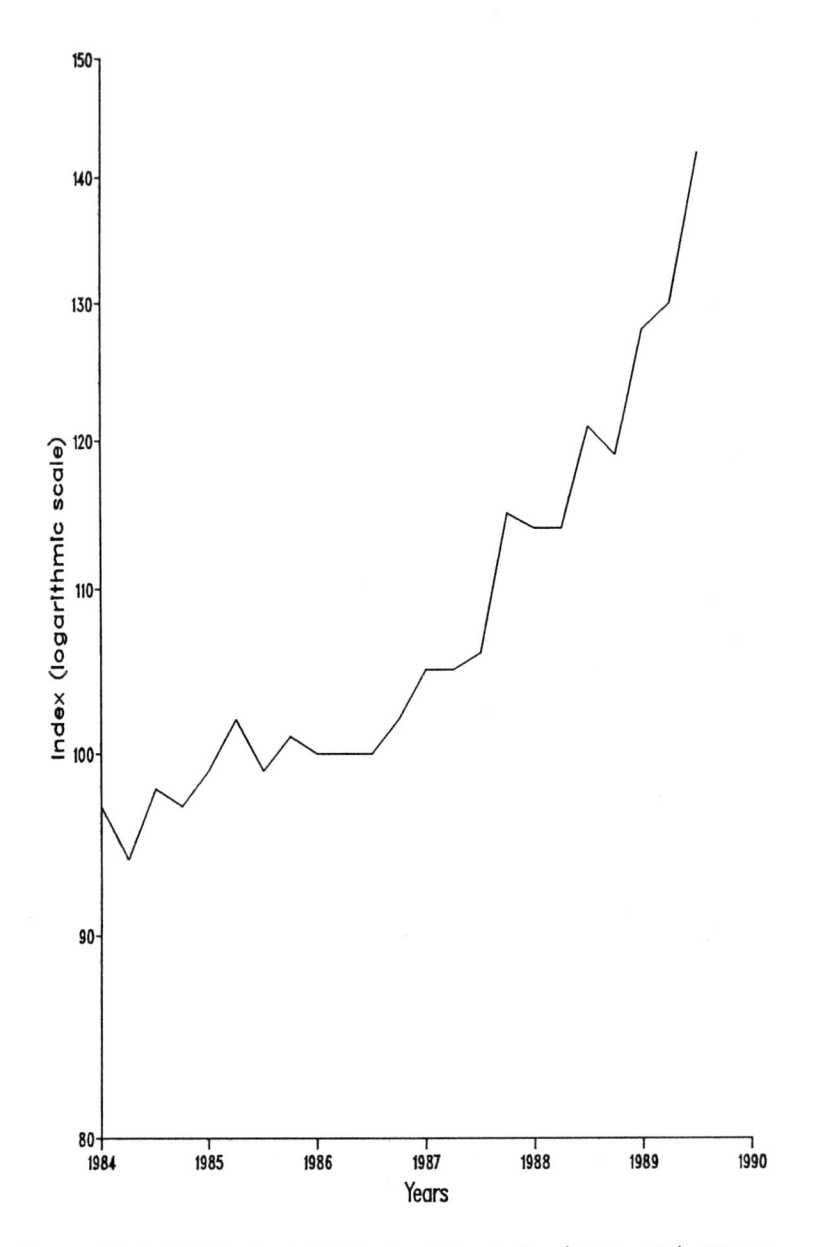

Figure 5.3.5 BCIS Rest-of-UK Tender Price Index (1985 = 100), 1984 Q1 – 1989 Q3

Table 5.3.6 BCIS Public Sector Tender Price Index

Base: 1985 = 100

Year	Q1	Q2	Q3	Q4	Average
1984	95	92	93	96	94
1985	99	105	99	100	101
1986	99	101	99	104	101
1987	105	104	109	118	109
1988	118	121	128	120	122
1989	132	136**	139**		
1990					
1991					
1992					
1993					
1994					
1995					

Source: BCIS.

** Provisional.

Table 5.3.7 BCIS Housing Tender Price Index

Base: 1985 = 100

Year	Q1	Q2	Q3	Q4	Average
1984	91	97	94	97	95
1985	98	101	99	103	100
1986	100	102	103	109	104
1987	107	109	108	117	110
1988	119	117	130	127	123
1989	132**	136**	132**		
1990					
1991					
1992					
1993					
1994					
1995					

Source: BCIS.

** Provisional.

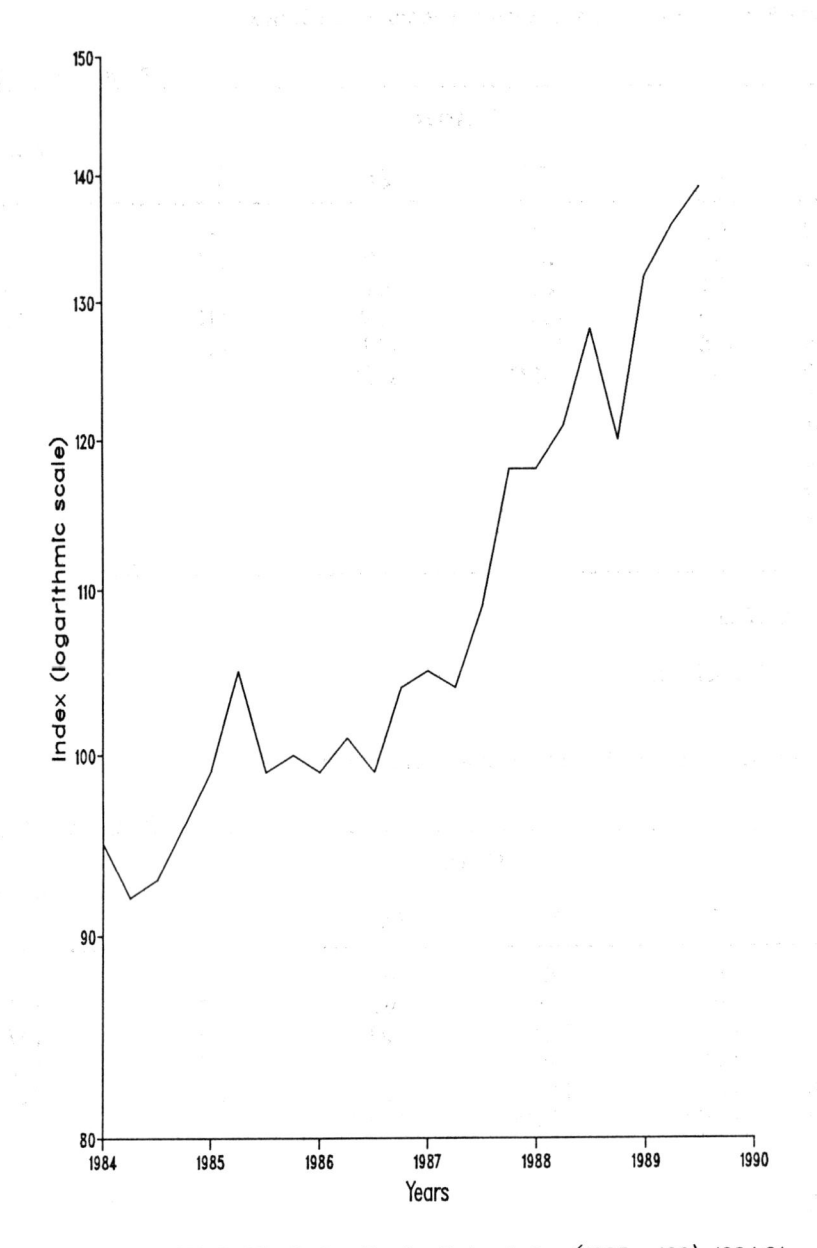

Figure 5.3.6 BCIS Public Sector Tender Price Index (1985 = 100), 1984 Q1 – 1989 Q3

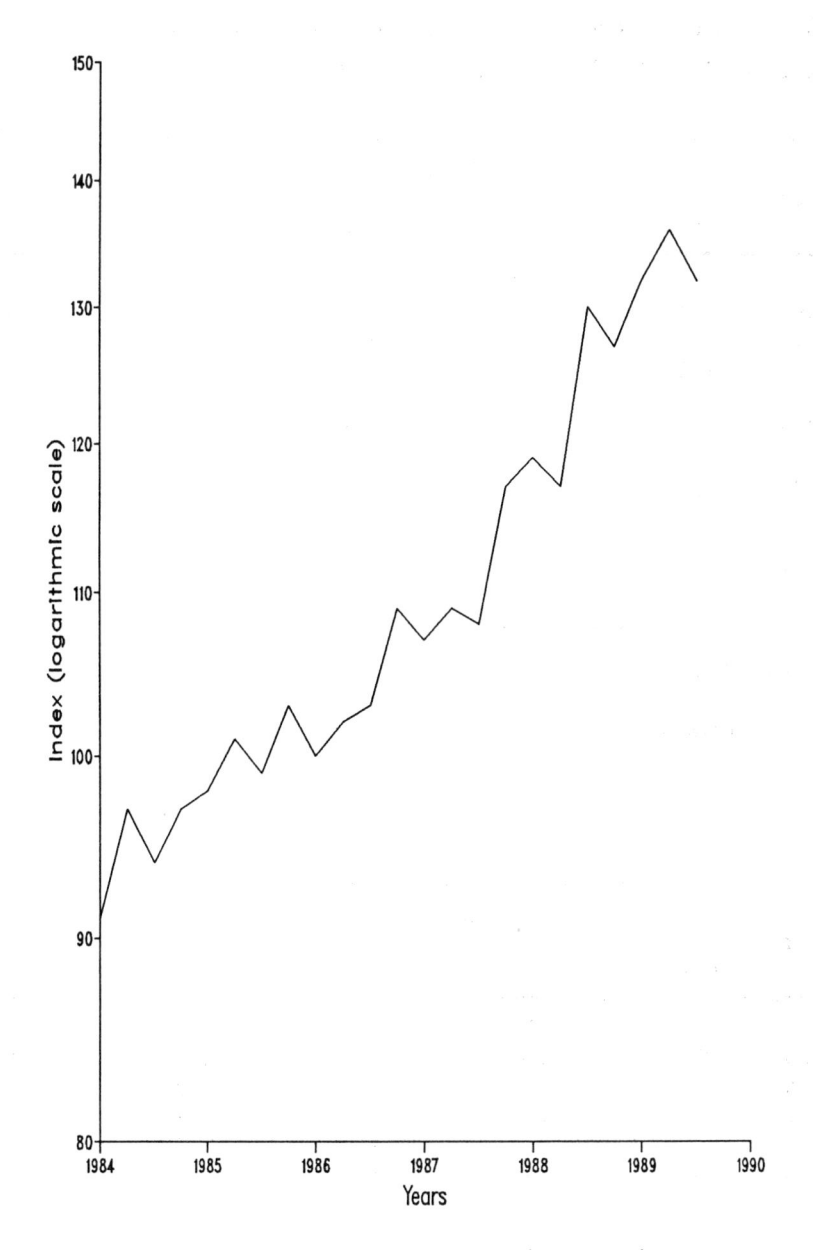

Figure 5.3.7 BCIS Housing Tender Price Index (1985 = 100), 1984 Q1 – 1989 Q3

Table 5.3.8 BCIS Private Sector Tender Price Index

Base: 1985 = 100

Year	Q1	Q2	Q3	Q4	Average
1984	91	97	104	94	97
1985	98	101	101	100	100
1986	101	103	105	104	103
1987	110	108	112	120	113
1988	124	124	129	130	127
1989	135	132**	138**		
1990					
1991					
1992					
1993					
1994					
1995					

Source: BCIS.

** Provisional.

Table 5.3.9 BCIS Private Commercial Tender Price Index

Base: 1985 = 100

Year	Q1	Q2	Q3	Q4	Average
1984	90	97	107	95	97
1985	97	102	101	102	100
1986	99	107	109	110	106
1987	111	108	112	124	114
1988	123	126	130	130	128
1989	134	136**	137**		
1990					
1991					
1992					
1993					
1994					
1995					

Source: BCIS.

** Provisional.

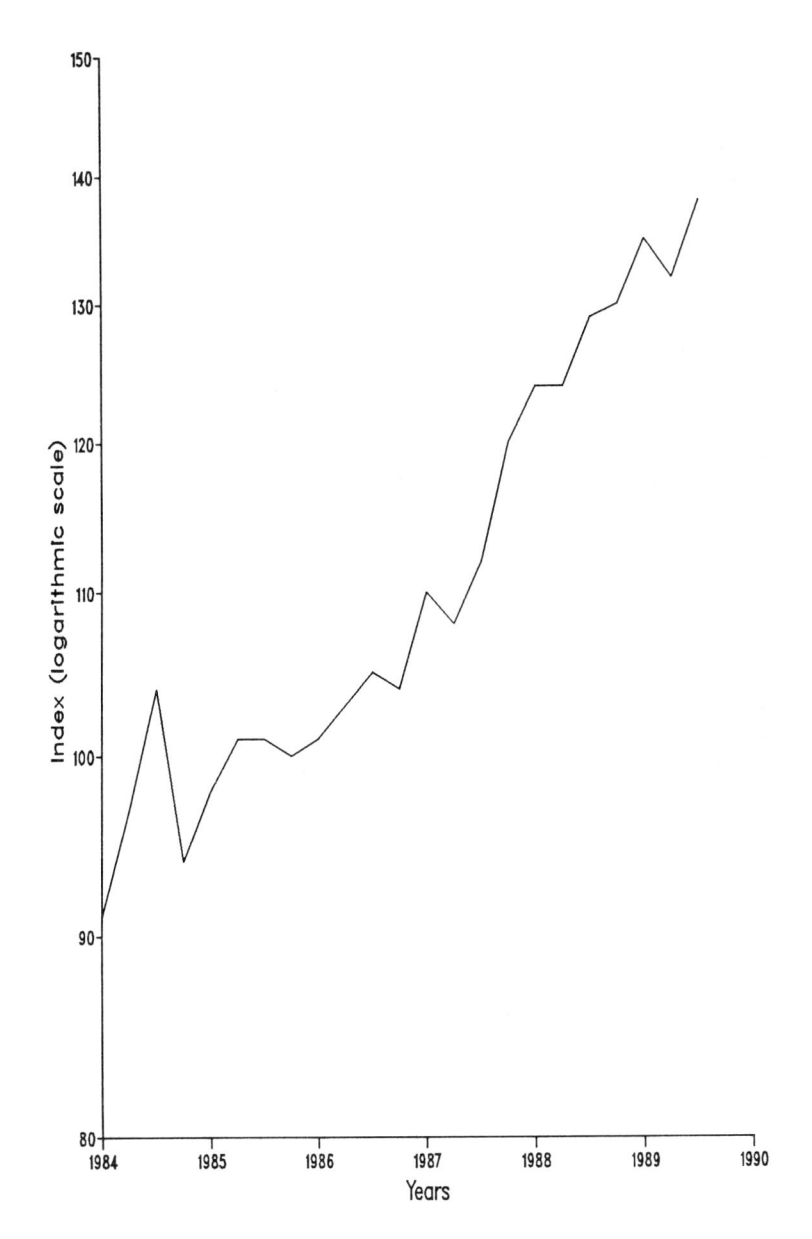

Figure 5.3.8 BCIS Private Sector Tender Price Index (1985 = 100), 1984 Q1 – 1989 Q3

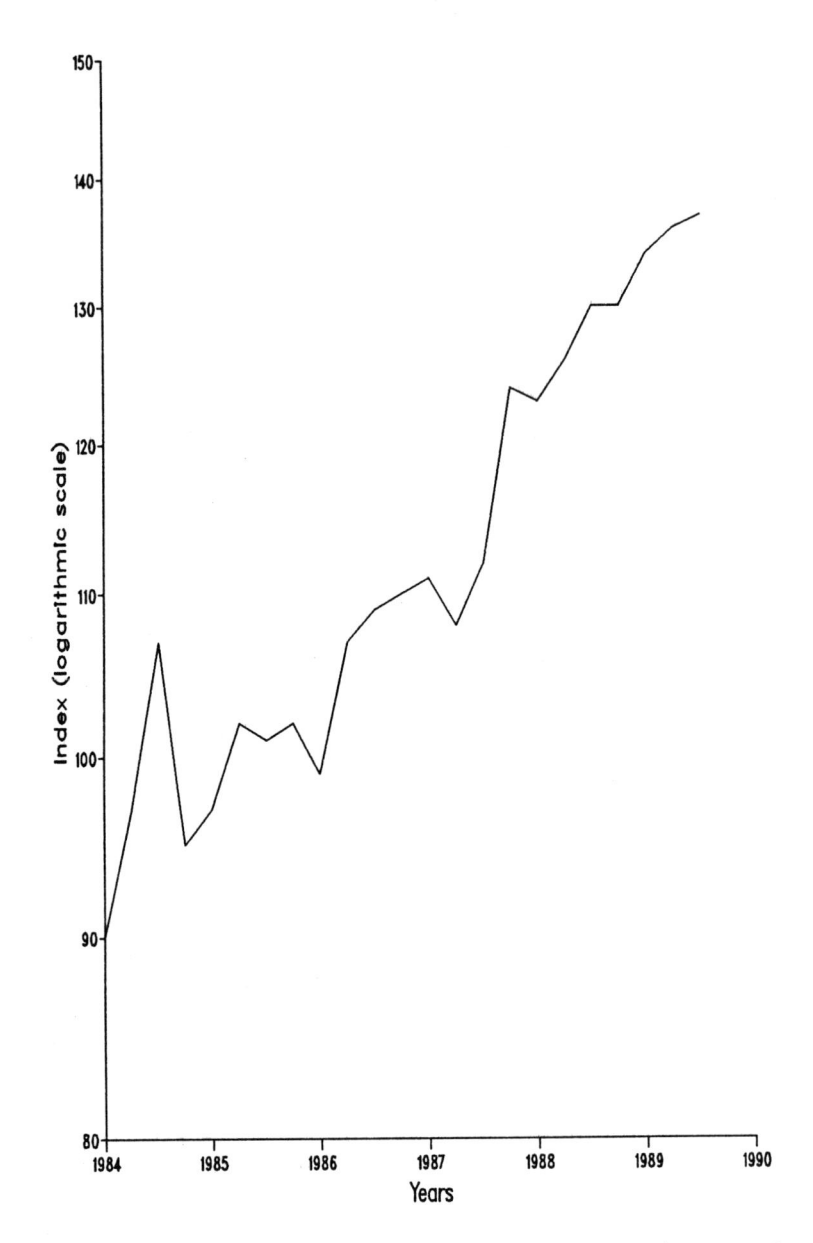

Figure 5.3.9 BCIS Private Commercial Tender Price Index (1985 = 100), 1984 Q1 – 1989 Q3

For Table 5.3.10, see overleaf.

Table 5.3.10 BCIS Private Industrial Tender Price Index

Base: 1985 = 100

Year	Q1	Q2	Q3	Q4	Average
1984	93	96	92	93	94
1985	103	101	99	98	100
1986	103	97	96	100	99
1987	106	107	109	109	108
1988	126	116	125	128	124
1989	130**	129**	142**		
1990					
1991					
1992					
1993					
1994					
1995					

Source: BCIS.

** Provisional.

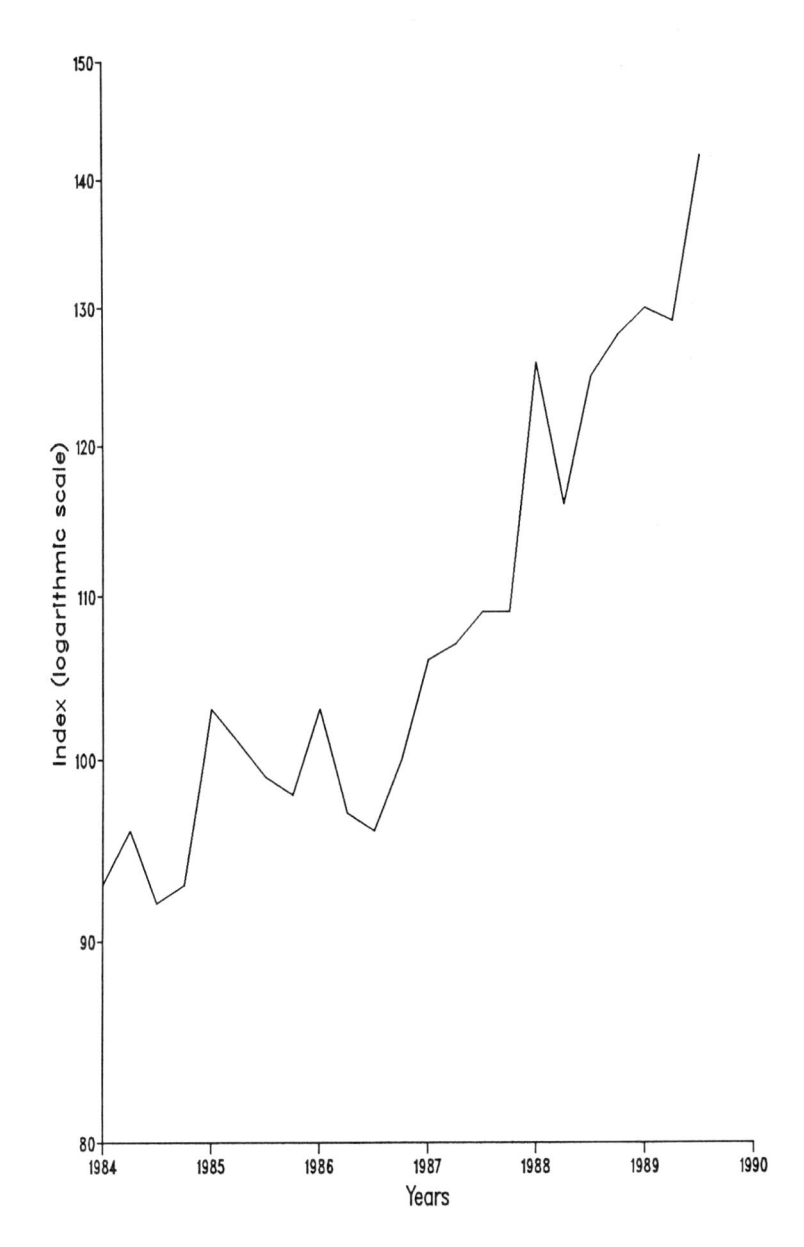

Figure 5.3.10 BCIS Private Industrial Tender Price Index (1985 = 100) 1984 Q1 – 1989 Q3

102 *Spon's handbook of construction cost and price indices*

5.4 DOE ROAD CONSTRUCTION TENDER PRICE INDICES

Type of Index

Tender price index (TPI)

Series: Coverage and Breakdowns

Analyses of new road construction in the public sector with breakdowns by type of contract and road as follows:

Series	*Table Reference*
All-in road construction TPI, 1970 to date	5.4.1
Firm price road construction TPI, 1975 to date	5.4.2
Variation of price road construction TPI, 1975 to date	5.4.3
Road construction TPI - trunk roads, 1975 to date	5.4.4
Road construction TPI - principal roads, 1975 to date	5.4.5

Base Dates and Period Covered

1975	from 1970 to 1979
1980	from 1975 to 1988
1985	from 1980 to date

The pre-1985 series are shown with 1985 = 100 in the tables below (converted arithmetically by the authors).

Frequency

Quarterly.

Geographical Coverage

Great Britain.

Type and Source of Data

Sample of selected tenders for new road construction in the public sector priced in competition with contract sums exceeding a defined level (currently £250,000).

Method of Compilation

The following is an extract from *Housing and Construction Statistics 1978-1988*.

'The overall index provides a measure of the change in tender prices for road construction in Great Britain. It is based on new contracts with a works cost of £250,000 or more. Before 1979 the lower limit was £100,000 in England and Wales and £25,000 in Scotland.

The index is produced by repricing, using a schedule of 1984/85 prices, after making an adjustment for preliminary and balancing items, the quantifiable items in bills of quantities for accepted tenders. The total adjusted cost of the quantifiable items is divided by their total adjusted cost over all contracts at 1984/85 prices, in order to calculate weighted average price relatives. An average calculated from a single quarter's contracts, often relatively few in number, would be over-sensitive to tender prices of individual large schemes, so each quarterly index value is the average price relative for contracts let in that quarter and the preceding and following quarters. Although in repricing tenders 1984/85 base prices are used, for publication the index is scaled to 1985 = 100. The above description of the derivation of the index refers to 1984 and later years. The index for years prior to 1984 is based on 1979/80 prices and scaled, 1984 being the link year. Sub-indices are produced for different road types and for types of contracts. Trunk motorways and Trunk roads in England are the responsibility of the Department of Transport whilst in Scotland and Wales the Scottish or Welsh Offices undertake this role. County Councils construct the principal roads including principal motorways.'

Commentary

The indices measure the change in tender price levels between one time period and another for road construction. Their main use, therefore, is in updating tenders or producing an estimate of current tender prices on the basis of historic cost information for road construction.

As the published indices represent general trends, their application in any particular use needs to take account of regional differences in price levels and possibly size of contract and any specific or unusual features.

Publications

(a) Data Source

DOE, *Housing and Construction Statistics* (HMSO).

(b) Description of Methodology

Details of the methodology are given in *Housing and Construction Statistics, 1978-1988*, HMSO, London.

Table 5.4.1 DOE All-In Road Construction Tender Price Index

Base: 1985 = 100*

Year	Q1	Q2	Q3	Q4	Average
1970	20	20	21	21	21
1971	22	22	22	22	22
1972	23	23	23	23	23
1973	25	26	28	32	28
1974	34	37	40	43	39
1975	44	44	43	41	43
1976	41	40	40	41	41
1977	45	47	50	53	49
1978	57	60	61	64	61
1979	67	72	76	88	76
1980	96	100	101	96	98
1981	94	91	89	87	90
1982	88	91	95	96	93
1983	97	95	94	95	94
1984	95	96	95	99	96
1985	98	100	97	102	100
1986	97	93	94	95	95
1987	101	106	107	111	106
1988	110	114	117	116	114
1989	130	125**	129**		
1990					
1991					
1992					
1993					
1994					
1995					

Source: DOE *Housing and Construction Statistics* (HMSO).

*   Indices from 1970 to 1979 inclusive converted arithmetically by or the authors from bases of 1975 = 100 or 1980 = 100 to 1985 = 100.

**   Provisional.

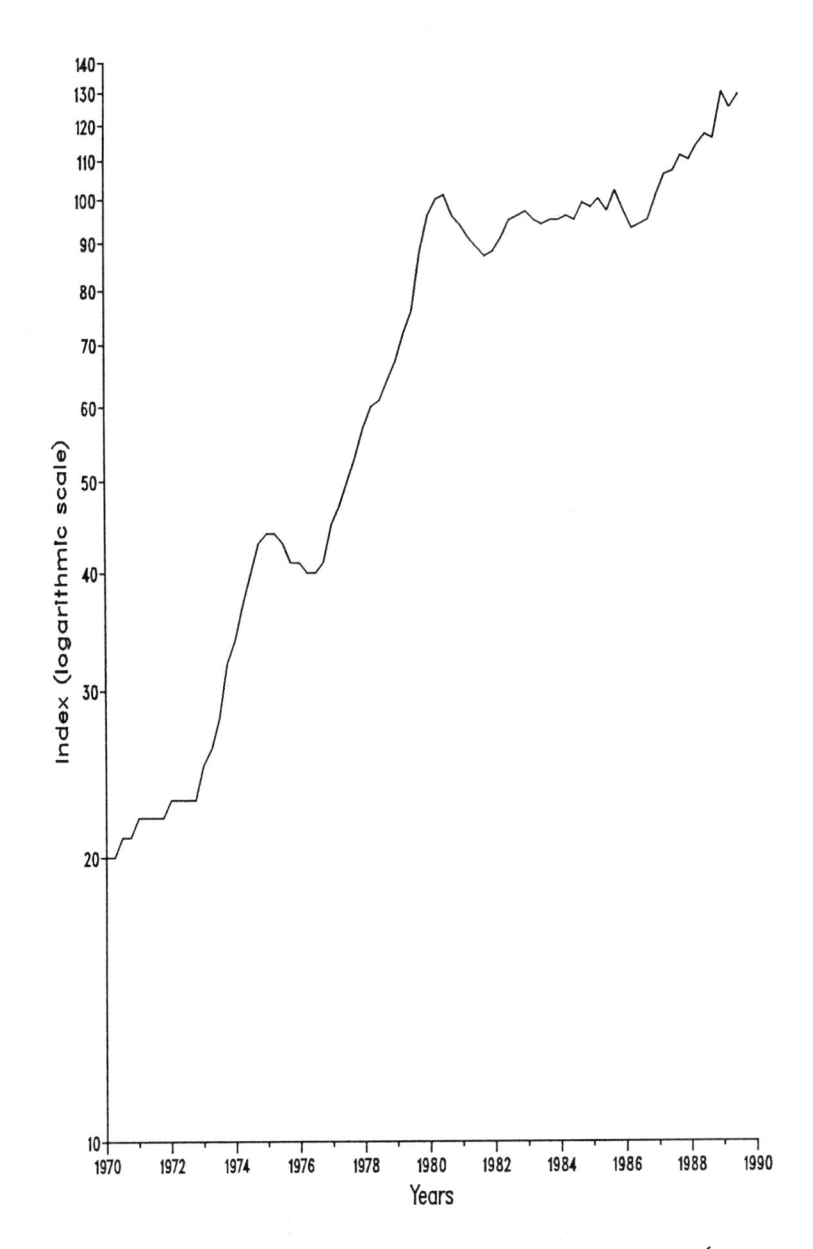

Figure 5.4.1 DOE All-in Road Construction Tender Price Index (1985 = 100) 1970 Q1 – 1989 Q3

Table 5.4.2 DOE Firm Price Road Construction Tender Price Index

Base: 1985 = 100*

Year	Q1	Q2	Q3	Q4	Average
1975	39	42	40	40	40
1976	41	45	45	47	45
1977	47	52	52	55	51
1978	56	59	59	63	59
1979	62	71	71	88	73
1980	95	100	95	94	96
1981	80	81	80	80	80
1982	90	85	101	96	93
1983	98	90	93	93	94
1984	99	99	99	103	100
1985	102	103	98	97	100
1986	94	93	95	95	94
1987	100	105	108	111	106
1988	113	112	116	110	113
1989	130	124**	127**		
1990					
1991					
1992					
1993					
1994					
1995					

Source: DOE *Housing and Construction Statistics* (HMSO).

*   Indices from 1975 to 1979 inclusive converted arithmetically by the authors from base of 1980 = 100 to 1985 = 100.

** Provisional.

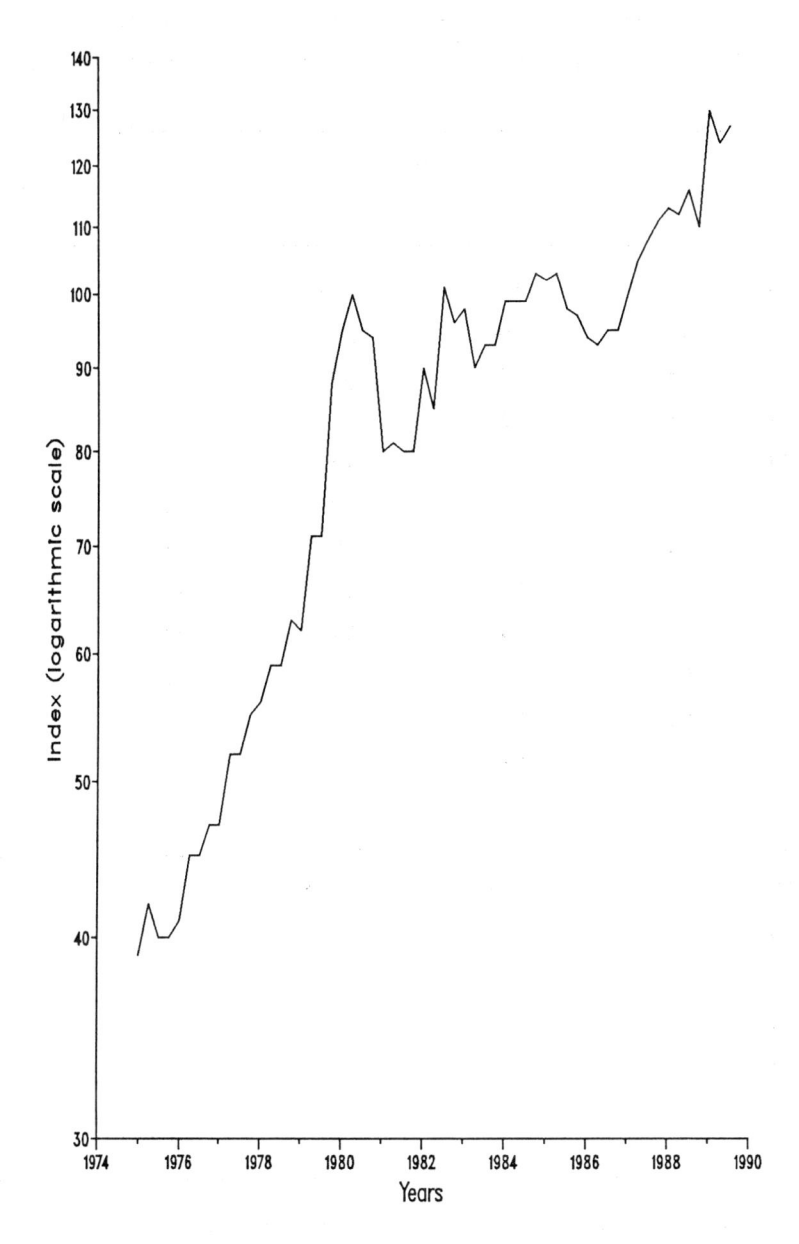

Figure 5.4.2 DOE Firm Price Road Construction Tender Price Index (1985 = 100), 1975 Q1 – 1989 Q3

Table 5.4.3 DOE Variation of Price Road Construction Tender Price Index

Base: 1985 = 100*

Year	Q1	Q2	Q3	Q4	Average
1975	45	44	43	42	43
1976	42	40	39	40	40
1977	45	47	50	53	49
1978	58	60	61	63	60
1979	67	72	75	87	75
1980	95	100	101	96	98
1981	93	90	89	86	90
1982	87	91	94	96	92
1983	96	94	93	94	94
1984	94	94	92	95	94
1985	94	98	99	109	100
1986	103	91	89	91	94
1987	100	103	99	109	103
1988	107	114	116	123	115
1989	124	133**	137**		
1990					
1991					
1992					
1993					
1994					
1995					

Source: DOE *Housing and Construction Statistics* (HMSO).

* Indices from 1975 to 1979 inclusive converted arithmetically by the authors from base of 1980 = 100 to 1985 = 100.

** Provisional.

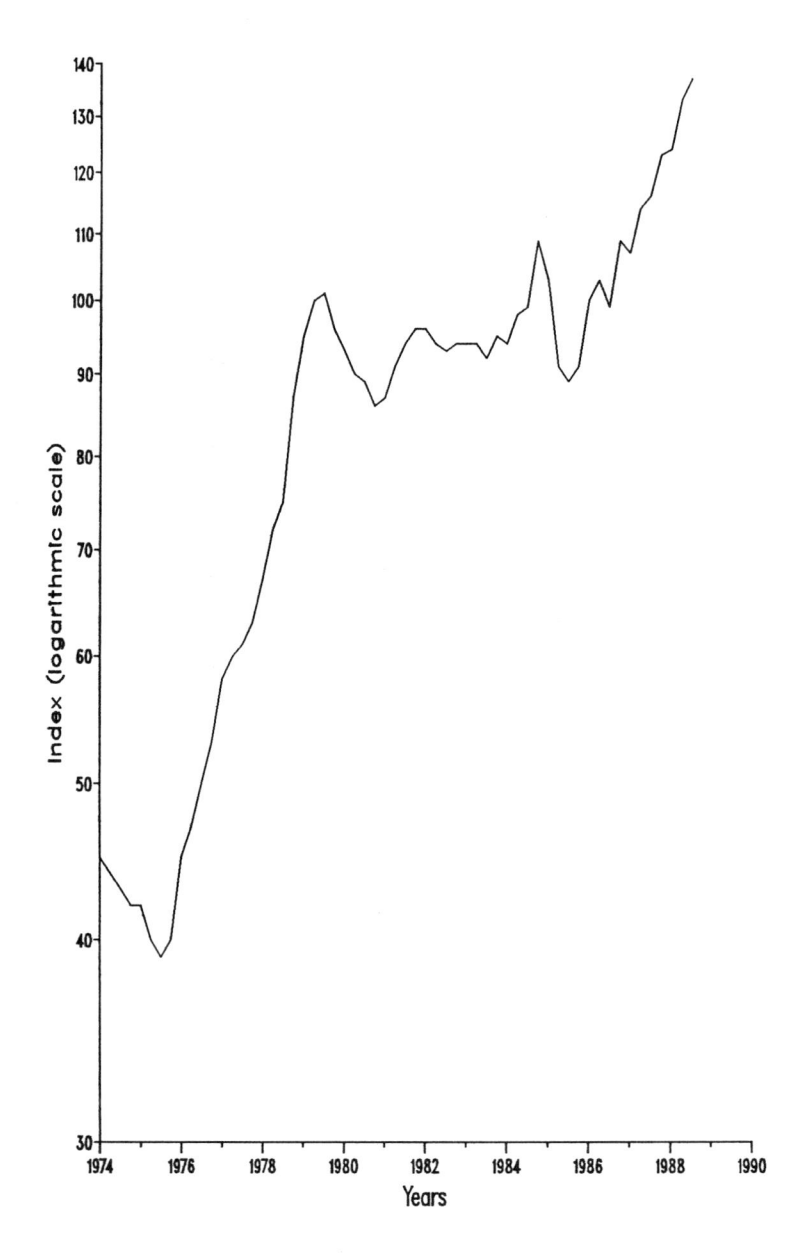

Figure 5.4.3 DOE Variation of Price Road Construction Tender Price Index (1985 = 100), 1975 Q1 – 1989 Q3

Table 5.4.4 DOE Road Construction Tender Price Index - Trunk Roads

Base: 1985 = 100*

Year	Q1	Q2	Q3	Q4	Average
1975	42	41	41	41	41
1976	41	40	39	41	40
1977	46	47	50	53	49
1978	57	58	63	68	61
1979	68	69	70	82	73
1980	81	93	94	97	91
1981	95	91	89	85	90
1982	88	91	93	95	92
1983	95	91	93	94	93
1984	97	95	97	100	97
1985	98	101	98	103	100
1986	95	92	94	95	94
1987	104	106	110	107	107
1988	103	104	114	105	107
1989	116	109**	113**		
1990					
1991					
1992					
1993					
1994					
1995					

Source: DOE *Housing and Construction Statistics* (HMSO).

* Indices from 1975 to 1979 inclusive converted arithmetically by the authors from base of 1980 = 100 to 1985 = 100.

** Provisional.

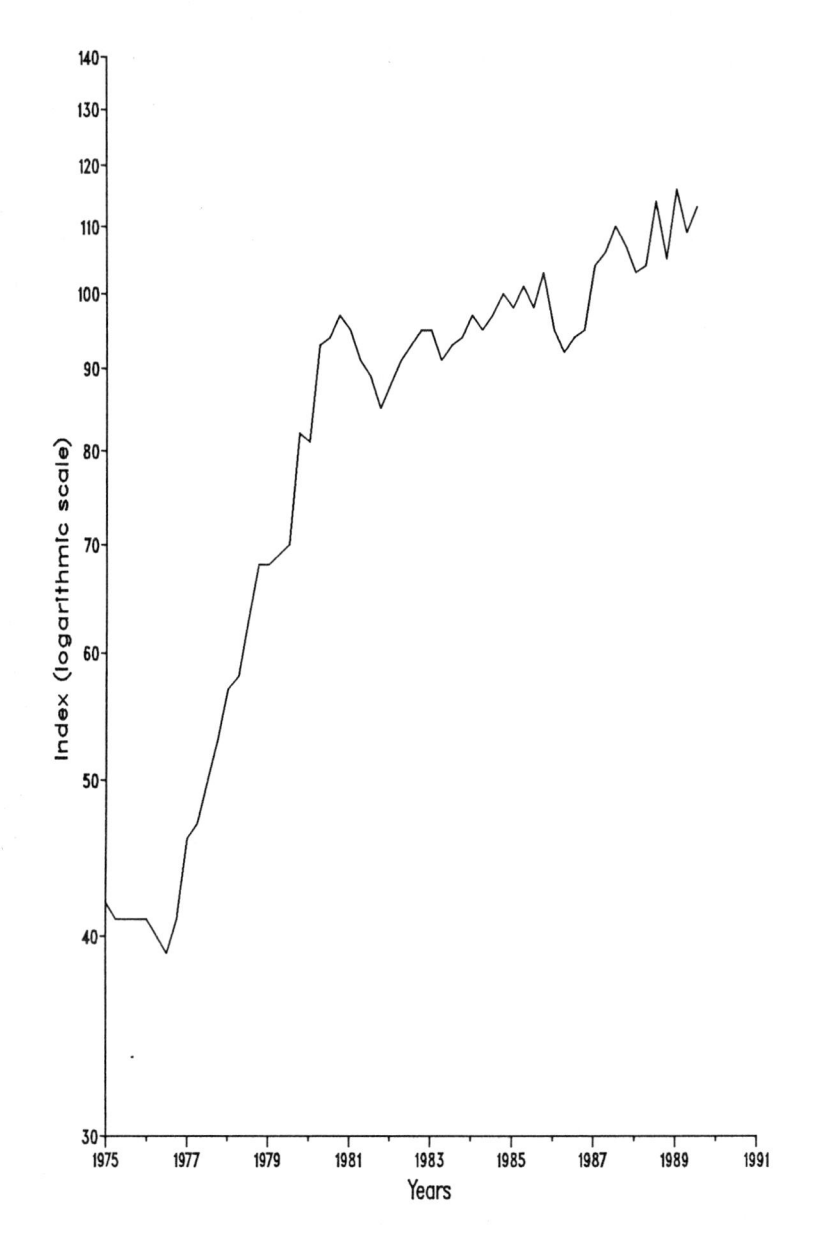

Figure 5.4.4 DOE Road Construction Tender Price Index – Trunk Roads – (1985 = 100), 1975 Q1 – 1989 Q3

Table 5.4.5 DOE Road Construction Tender Price Index - Principal Roads

Base: 1985 = 100*

Year	Q1	Q2	Q3	Q4	Average
1975	40	46	45	44	44
1976	44	46	49	51	48
1977	49	54	55	57	54
1978	58	69	66	69	65
1979	67	74	93	106	85
1980	101	98	92	91	96
1981	84	85	88	87	86
1982	83	80	79	83	81
1983	85	90	85	90	87
1984	88	95	88	93	91
1985	98	101	103	98	100
1986	100	100	93	93	97
1987	92	96	97	103	97
1988	113	111	120	124	117
1989	141	134**	132**		
1990					
1991					
1992					
1993					
1994					
1995					

Source: DOE *Housing and Construction Statistics* (HMSO).

* Indices from 1975 to 1979 inclusive converted arithmetically by the authors from base of 1980 = 100 to 1985 = 100.

** Provisional.

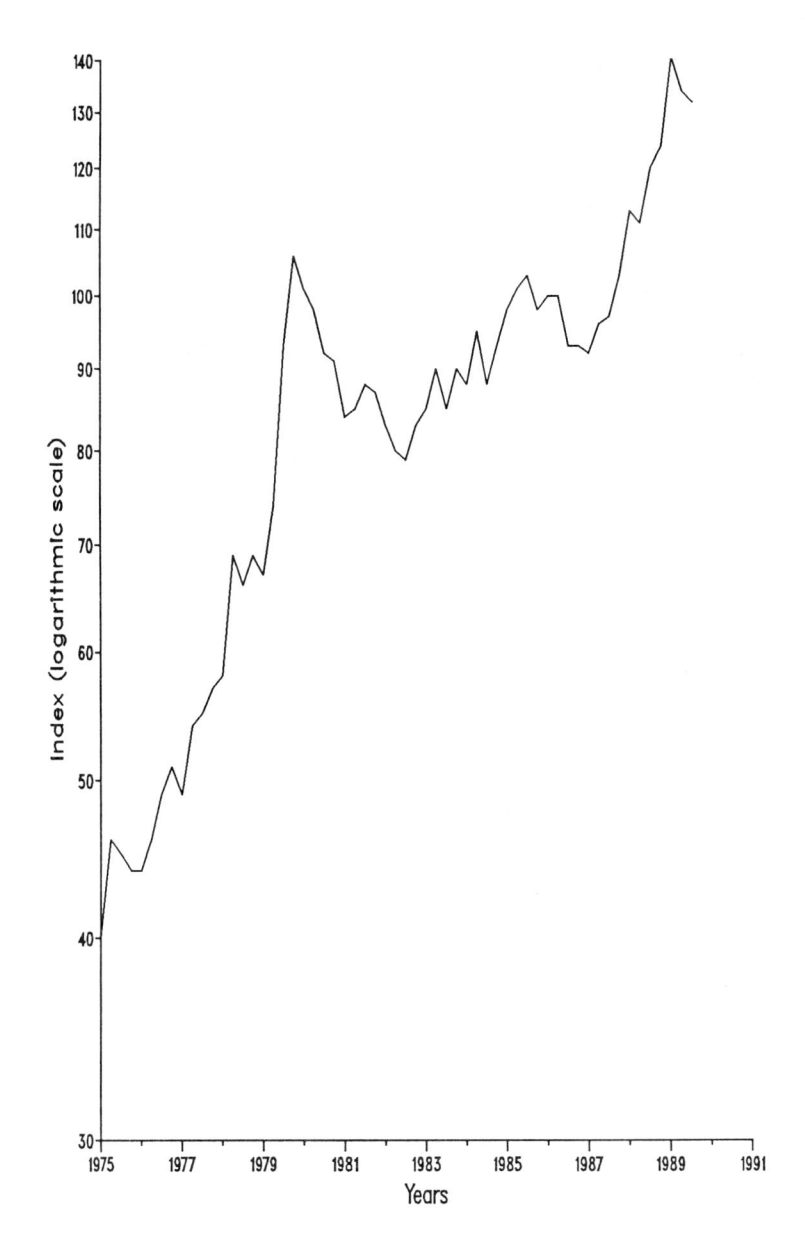

Figure 5.4.5 DOE Road Construction Tender Price Index – Principal Roads – (1985 = 100), 1975 Q1 – 1989 Q3

5.5 DOE PRICE INDEX OF PUBLIC SECTOR HOUSEBUILDING

Type of Index

Tender-related price index.

Series: Coverage and Breakdowns

Analyses of new house building in the public sector with breakdowns by type of contract as follows:

Series	Table Reference
All-in TPI, 1964 to date	5.5.1
Firm price TPI, 1975 to date	5.5.2
Variation of price TPI, 1975 to date	5.5.3

Base Dates and Period Covered

1970	from 1966 to 1978
1975	from 1975 to 1983
1980	from 1975 to 1988
1985	from 1980 to date

The pre-1985 series are shown with 1985 = 100 in the tables below (converted arithmetically by either the DOE or the authors).

Frequency

Quarterly.

Geographical Coverage

England and Wales.

Type and Source of Data

Accepted tenders for traditionally built one- and two- storey local authority housing and, latterly, three- and four- storey blocks of dwellings.

Method of Compilation

The following is an extract from *Housing and Construction Statistics 1978-1988*.

'This index provides a measure of the change in tender prices for the construction of public sector housing in England and Wales. It has superseded, from 1979, the price index of local authority housebuilding (PILAH) which was produced until the fourth quarter of 1978. PILAH was based on tenders accepted for traditionally built one- and two-storey local authority housing, excluding Greater London. As a result of a review the coverage of PILAH was extended to also include contracts let by London Boroughs, Greater London Council, new towns and housing associations and to include dwellings in three- and four-storey blocks, so that the new index is an index for the whole of public sector housebuilding in England and Wales. The base prices and weights used in PILAH have been revised. The methodology of the new index is the same as that of PILAH. The price data are extracted from the bills of quantities of successful tenders for dwellings in blocks up to four storeys built by traditional methods. Within this field the price changes are measured from the prices of 23 items which occur in most such bills of quantities. As part of the review of PILAH there have been some revisions to the definitions of these items. Each item is selected to represent all the work in a particular trade section so that price movements of other work in a trade section are assumed to be broadly similar to those of the representative item. Table A shows the list of items. The index is compiled as a Laspeyres price index in which a weighted arithmetic average of the price relatives for each item in the current quarter, relative to the base year is taken. Separate indices are produced for Variation of Price and Firm Price contracts based on 1980. Price indices from 1980 are calculated on a national basis instead of the weighted regional basis used in earlier years, as a review of the regional weights indicated that the regional analysis did not exhibit consistent price differentials. Values of the price index of public sector housebuilding and PILAH were calculated for the fourth quarter of 1978 so that the new series could be linked to the PILAH series. This linking was done by applying the rise between the fourth quarter of 1978 and the first quarter of 1979 of the price index of local authority housebuilding to the PILAH value for the fourth quarter of 1978 to give a value for the first quarter of 1979. Subsequent quarterly movements are calculated from the values of the new index.'

Table A Weights used for trades or operations

Item	Weights used for 1985 base %	Item	Weights used for 1985 base %
1 Excavation & hardcore	3	13 Copper plumbing	2
2 Concrete work	11	14 Hot & cold water tanks	1
3 Blockwork - general	13	15 Sanitary fittings	3
4 Brickwork - facing	7	16 Central heating	8
5 Partitions	0	17 Electrical installation	6
6 Roof tiling	5	18 Plastering - walls	5
7 Carpentry	6	19 Ground floor covering	3
8 First floor boarding	0	20 Plastering - ceilings	2
9 Manufactured joinery - doors	6	21 Glazing	1
10 Windows	4	22 Wall and ceiling paint	2
11 General joinery	8	23 Oil painting	3
12 Soil and vent piping	1		

Values of the index for 1984 and later years are based on 1985 = 100, the items and their weights being as shown in Table A. Values of the index for earlier years are obtained as above and scaled so that 1984 is the link year.

Treatment of missing items

Although the 23 items are chosen to represent a typical building scheme there are some schemes in which one or more of the items are not in fact used. When this occurs a price relative is inserted for the scheme so as to produce a balanced index. The price value chosen is the average for this item in the remainder of the schemes during that quarter. Occasionally it may happen that a particular item or items is not represented in any scheme during that quarter. When this happens it is assumed that the price of the item changes in the same manner as the (weighted) index for the remaining items. An article in *Statistical News* (HMSO) for August 1973 discusses the methodology of PILAH. It also contains some historical notes.'

Commentary

As will be noted from the foregoing, schemes in Greater London were specifically excluded until the first quarter of 1979. This was because of the relatively small amount of low-rise traditional house building in Greater London. The change in the index from the first quarter of 1979 has overcome this problem.

The indices measure the change in tender price levels between one time period and another. Their main use, therefore, is in updating tenders or producing an estimate of current tender prices on the basis of historic cost information. As the published indices represent general trends, their application in any particular use needs to take account of regional differences in price levels (the PSA produce a guide to the influence of locational factors from time to time as a study), and possibly size of contract and any specific or unusual features.

Publications

(a) Data Source

DOE, *Housing and Construction Statistics* (HMSO).

(b) Description of Methodology

Details of the methodology are given in *Housing and Construction Statistics 1978-1988*, HMSO, London and the methodology of PILAH is discussed in an article in *Statistical News*, HMSO, London, for August 1973.

Table 5.1 DOE Price Index of Public Sector Housebuilding - All-In

Base: 1985 = 100*

Year	Q1	Q2	Q3	Q4	Average
1970	18	18	18	19	18
1971	20	21	21	22	21
1972	23	24	25	28	25
1973	30	33	36	39	35
1974	39	39	41	41	40
1975	42	42	43	45	43
1976	44	46	46	46	46
1977	50	50	52	53	51
1978	55	57	58	62	58
1979	62	66	71	74	68
1980	81	84	87	88	84
1981	87	84	84	85	85
1982	88	87	88	89	88
1983	91	94	96	94	94
1984	96	96	96	97	96
1985	97	99	101	104	100
1986	105	103	104	103	104
1987	109	112	114	117	113
1988	120	124	125	132	125
1989	134	138	136	137**	136**
1990					
1991					
1992					
1993					
1994					
1995					

Source: DOE *Housing and Construction Statistics* (HMSO).

*   Indices from 1970 to 1979 inclusive converted arithmetically by the authors from base of 1975 = 100 to 1985 = 100.

**   Provisional.

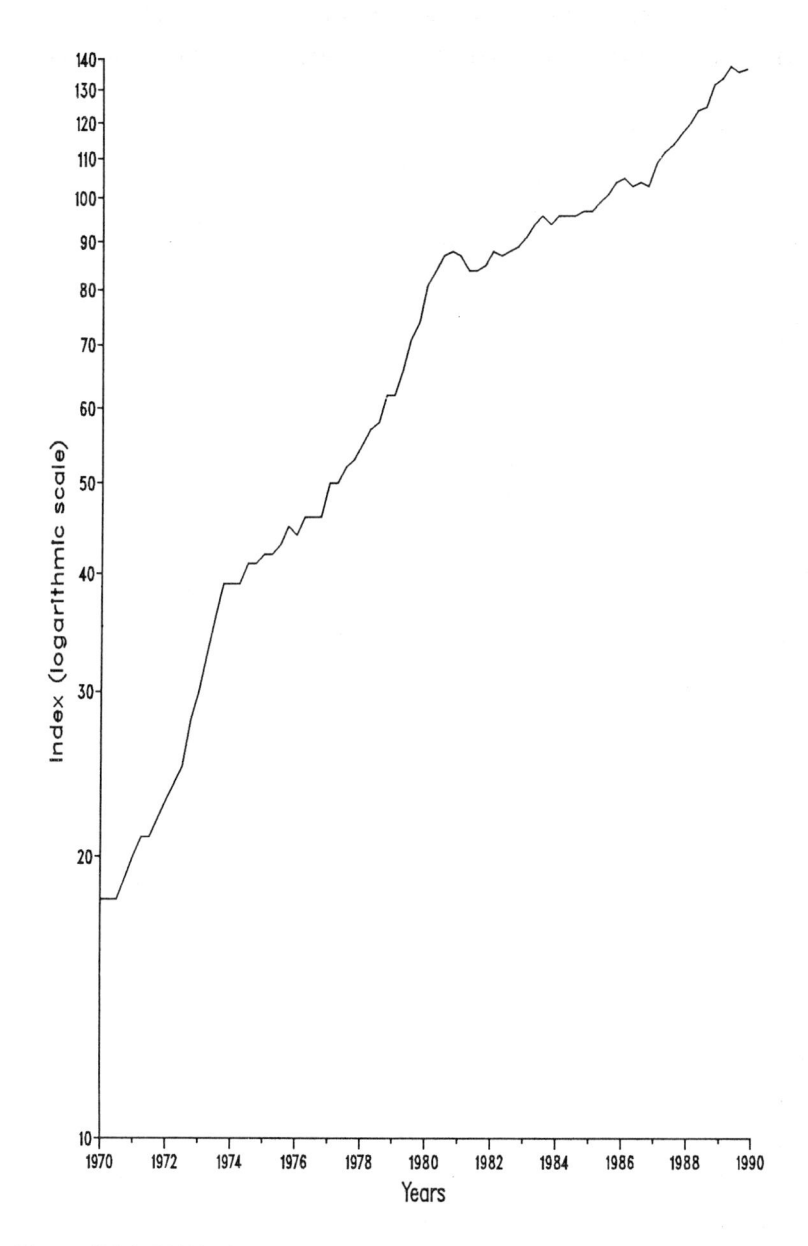

Figure 5.5.1 DOE Price Index of Public Sector Housebuilding – All-in – (1985 – 100), 1970 Q1 – 1989 Q4

Table 5.5.2 DOE Price Index of Public Sector Housebuilding - Firm Price

Base: 1985 = 100*

Year	Q1	Q2	Q3	Q4	Average
1975	42	41	44	44	42
1976	44	45	46	46	45
1977	49	50	51	54	51
1978	54	60	59	62	59
1979	61	67	72	74	68
1980	82	86	88	87	86
1981	88	83	86	86	86
1982	88	88	88	89	88
1983	92	94	96	95	94
1984	95	97	97	99	97
1985	98	99	100	103	100
1986	104	102	103	102	103
1987	108	111	113	115	112
1988	119	122	122	130	123
1989	132	136	134	135**	134**
1990					
1991					
1992					
1993					
1994					
1995					

Source: DOE *Housing and Construction Statistics* (HMSO).

*   Indices from 1975 to 1979 inclusive converted arithmetically by the authors from base of 1975 = 100 to 1985 = 100.

** Provisional.

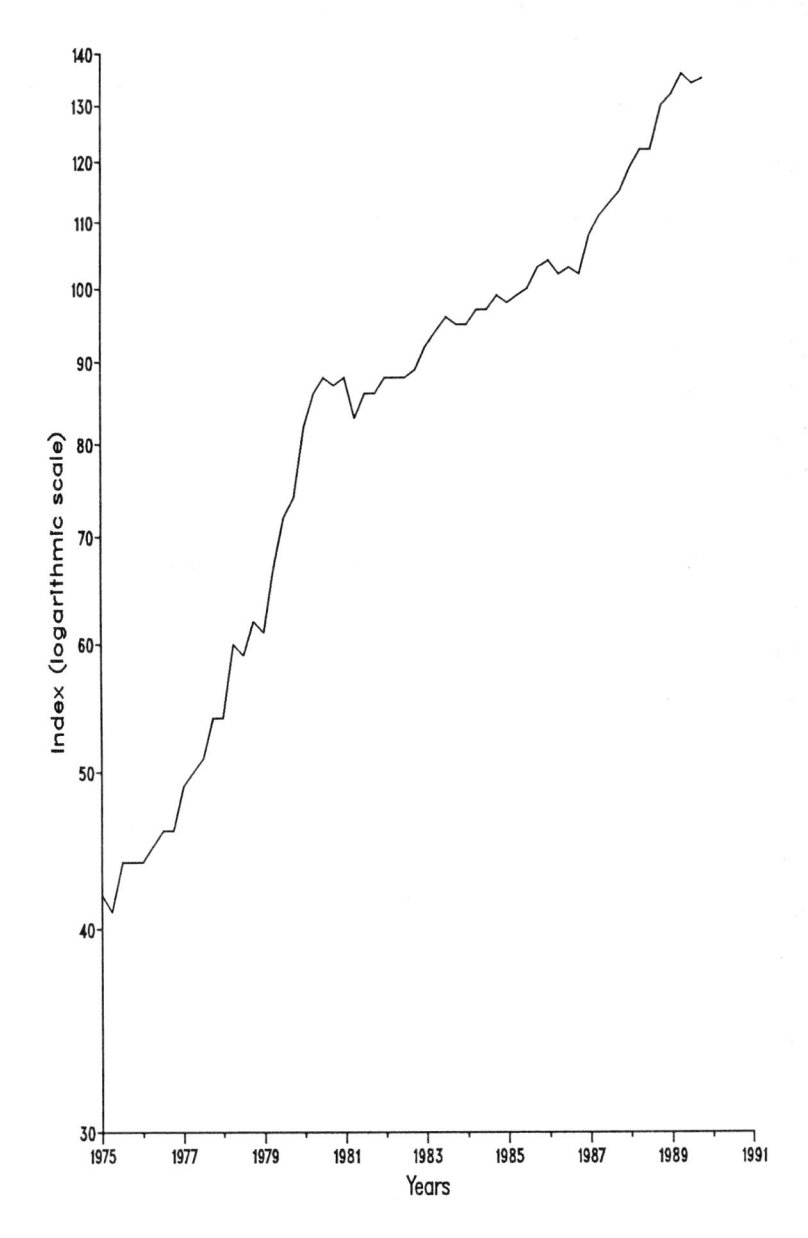

Figure 5.5.2 DOE Price Index of Public Sector Housebuilding – Firm Price – (1985 = 100), 1975 Q1 – 1989 Q4

Table 5.5.3 DOE Price Index of Public Sector Housebuilding - Variation of Price

Base: 1985 = 100*

Year	Q1	Q2	Q3	Q4	Average
1975	42	43	43	44	43
1976	44	44	45	47	45
1977	49	49	51	52	51
1978	54	56	57	62	57
1979	62	68	70	75	69
1980	81	86	87	90	86
1981	88	84	83	86	85
1982	88	88	88	89	88
1983	91	95	96	93	94
1984	95	95	95	94	95
1985	96	98	100	106	100
1986	110	103	106	105	106
1987	109	115	106	-	111
1988	-	-	-	-	-
1989	-	-	-	-	-
1990					
1991					
1992					
1993					
1994					
1995					

Source: DOE *Housing and Construction Statistics* (HMSO).

* Indices from 1975 to 1979 inclusive converted arithmetically by the authors from base of 1975 = 100 to 1985 = 100.

** Provisional.

- Too few contracts for a viable index to be produced.

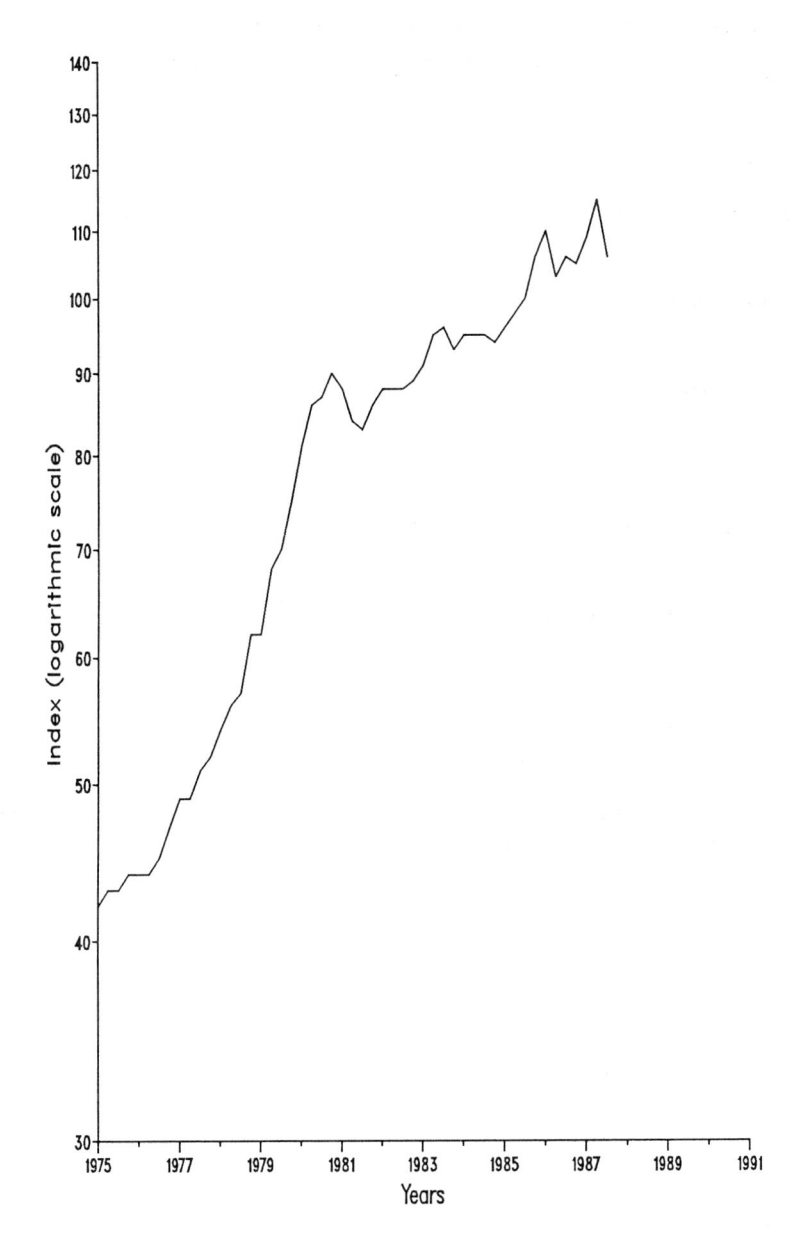

Figure 5.5.3 DOE Price Index of Public Sector Housebuilding – Variation of Price – (1985 = 100), 1975 Q1 – 1987 Q3

5.6 SCOTTISH OFFICE HOUSING TENDER PRICE INDEX

Type of Index

Tender price index (TPI).

Series: Coverage and Breakdowns

Analyses of new public sector house building by the Scottish Office Building Directorate, with no breakdowns by type of contract, as follows:

Series	Table Reference
All-in TPI, 1970 to date	5.6.1

Base Dates and Period Covered

1970	from 1970 to 1979
1975	from 1974 to 1983
1980	from 1980 to 1987
1985	from 1980 to date

The pre-1985 series are shown with 1985 = 100 in the tables below (converted arithmetically by the authors).

Frequency

Quarterly.

Geographical Coverage

Scottish mainland.

Type and Source of Data

Sample of tenders for traditionally built one- to four-storey public sector housing.

Method of Compilation

The index is compiled by the Scottish Office Building Directorate. It measures the movement of prices in competitive tenders for one- to four-storey public sector housing throughout the Scottish mainland. Up to and including 1983 Q1 the quarterly index was the mean of the individual project indices. From 1983 Q2 the index has been constructed using a statistical method to insulate against the effects of erratic short term changes without interfering with the response to genuine changes in price levels.

Individual project indices are calculated using a method similar to that used to compile the PSA QSSD index of building tender prices (see section 5.2), and the BCIS tender price indices (see section 5.3).

Commentary

The index measures the change in tender price levels for Scottish public sector housing between one time period and another. Its main use, therefore, is in updating tenders or producing an estimate of current tender prices on the basis of historic cost information for housing in Scotland. As the published index represents general trends, its application in any particular use needs to take account of regional differences in price levels and possibly size of contract and any specific or unusual features. It should also be noted that there is no differential between type of contract in this index.

Publications

(a) Data Source

Statistical Bulletins, quarterly, Housing Statistics Unit, Scottish Development Department, available from Scottish Office Library, Publications Sales, Room 2/65, New St Andrews House, Edinburgh, EH1 3TG.

(b) Description of Methodology

The construction, scope and uses of the index are as described above.

Table 5.6.1   Scottish Office Housing Tender Price Index

Base: 1985 = 100*

Year	Q1	Q2	Q3	Q4	Average
1970	19	19	19	19	19
1971	19	20	20	21	20
1972	22	22	24	26	24
1973	28	31	37	39	34
1974	42	43	43	46	44
1975	46	48	46	44	46
1976	46	47	47	47	47
1977	48	49	49	52	50
1978	51	56	58	59	56
1979	63	66	71	74	69
1980	82	86	82	84	84
1981	88	84	78	83	83
1982	82	87	87	84	85
1983	88	89	93	95	91
1984	95	99	99	104	99
1985	101	98	100	101	101
1986	105	101	102	103	103
1987	102	101	108	111	106
1988	114	112	117	124	117
1989	128	134	136	137**	134**
1990					
1991					

Source: Scottish Office

*   Indices from 1970 to 1984 inclusive converted arithmetically by the authors from base of 1970 = 100, 1975 = 100 and 1980 = 100 to 1985 = 100.

**   Provisional.

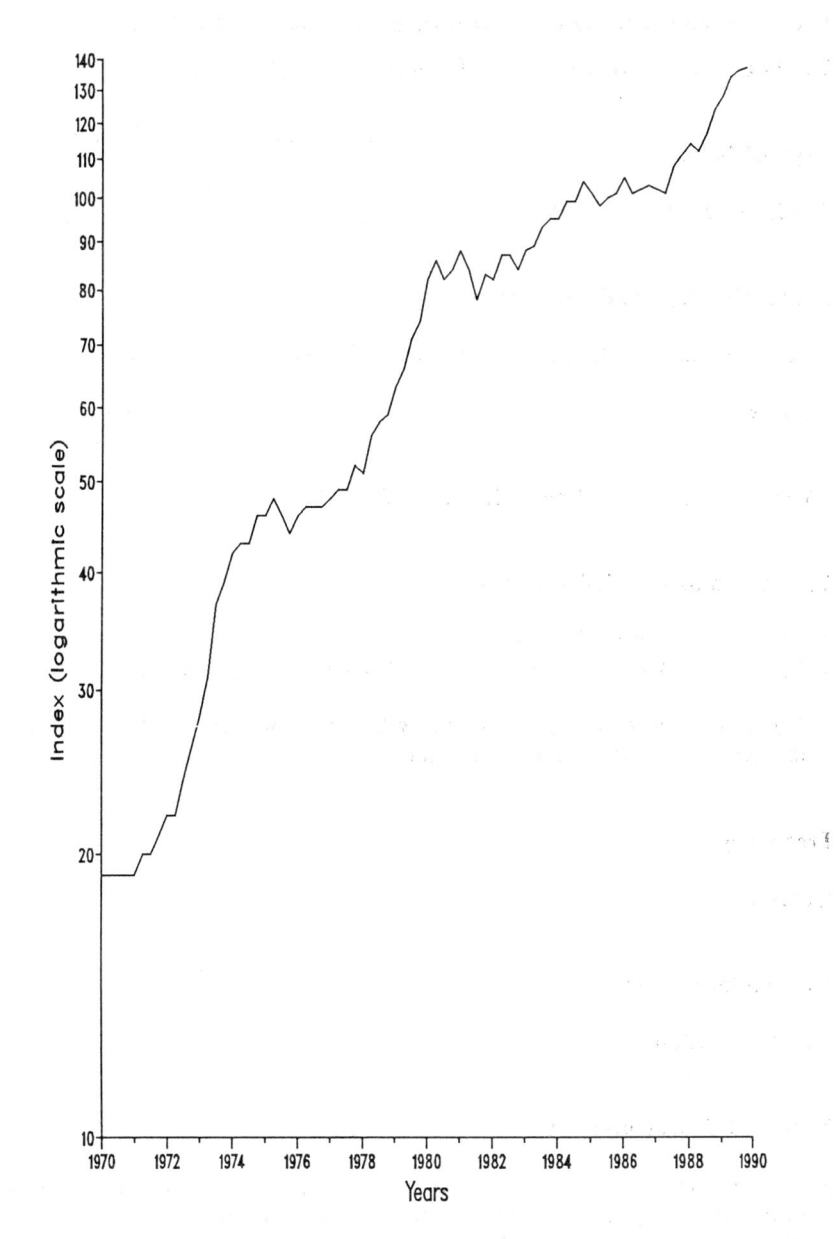

Figure 5.6.1 Scottish Office Housing Tender Price Index (1985 = 100) 1970 Q1 – 1989 Q4

5.7 D L & E TENDER PRICE INDEX

Type of Index

Tender price index (TPI).

Series: Coverage and Breakdowns

Analysis of new building in the public and private sectors as follows:

Series	Table Reference
Variation of Price, Greater London TPI, 1966 to date	5.7.1

Base Dates and Period Covered

1970	from 1966 to 1980.
1976	from 1976 to date with forecasts two years ahead.

The pre-1976 series are shown with 1976 = 100 in the tables below (converted arithmetically by the authors).

Frequency

Quarterly.

Geographical Coverage

Greater London.

Type and Source of Data

Sample of selected tenders for new building work (excluding work of a mainly civil engineering nature, mechanical and electrical work, complex works of alterations and extensions and negotiated contracts) in the public and private sectors priced in competition with contract sums exceeding a defined level (currently £300,000). Details of the tenders - priced bills of quantities - are generally obtained from within Davis, Langdon & Everest (D L & E), Chartered Quantity Surveyors.

Method of Compilation

The method used is that of re-pricing a large number of items selected from each trade or section of Bills of Quantities for current tenders on the basis of a standard schedule of rates. Preliminaries and other general charges are spread proportionately over each item of the Bill of Quantities. For each project the re-priced tender figure is then compared with the actual tender figure to produce a 'project index'. The index figures given are derived from the geometric mean of the project indices of each quarter's sample. The individual project index numbers are very variable and an attempt is made to base the published index on as many projects within each quarter as possible in order to reduce sampling error. As would be expected for an index applicable to a specific location, the number is often less than preferred by the BCIS or the PSA for inclusion in their indices, but this is offset by the greater sample of items included in the analysis.

Commentary

The indices measure the change in tender price levels between one time period and another. Their main use, therefore, is in updating tenders or producing an estimate of current tender prices on the basis of historic cost information. As the published indices represent general trends, their application in any particular use needs to take account of regional differences in price levels (D L & E produce a guide to the influence of locational factors in *Spon's Architects' and Builders' Price Book*), and possibly size of contract and any specific or unusual features.

Publications

(a) Data Source

Davis, Langdon and Everest, Princes House, 37-39 Kingsway, London WC2B 6TP.

Quarterly in *Building* magazine. *Building* also includes information on regional trends showing one year historic, and one year forecast, percentage changes in building tender prices for each region of Great Britain.

(b) Description of Methodology

Details of the methodology are available from within Davis, Langdon and Everest, but it is similar to that used by the BCIS, there are only very minor differences used within the calculations.

Spon's handbook of construction cost and price indices

Table 5.7.1 D L & E Tender Price Index

Base: 1976 = 100*

Year	Q1	Q2	Q3	Q4	Average
1966	34	34	34	34	34
1967	34	35	35	36	35
1968	36	36	37	37	36
1969	37	37	37	38	37
1970	39	41	43	43	42
1971	45	46	48	50	48
1972	55	57	61	67	60
1973	73	79	86	92	82
1974	97	100	99	97	98
1975	101	103	98	100	100
1976	97	98	102	103	100
1977	105	105	109	110	107
1978	113	116	126	139	124
1979	142	146	160	167	154
1980	179	200	192	188	190
1981	199	193	190	195	194
1982	191	188	195	195	192
1983	198	200	198	200	199
1984	205	206	214	215	210
1985	215	219	219	220	218
1986	221	226	234	234	229
1987	242	249	265	279	258
1988	289	299	321	328	309
1989	341	335	340	345	340
1990	330**				
1991					
1992					
1993					
1994					
1995					

Source: *Spon's Architects' and Builders' Price Book* and *Building*

*   Indices from 1966 to 1975 inclusive converted arithmetically by the authors from base 1970 = 100 to 1976 = 100.
** Provisional.

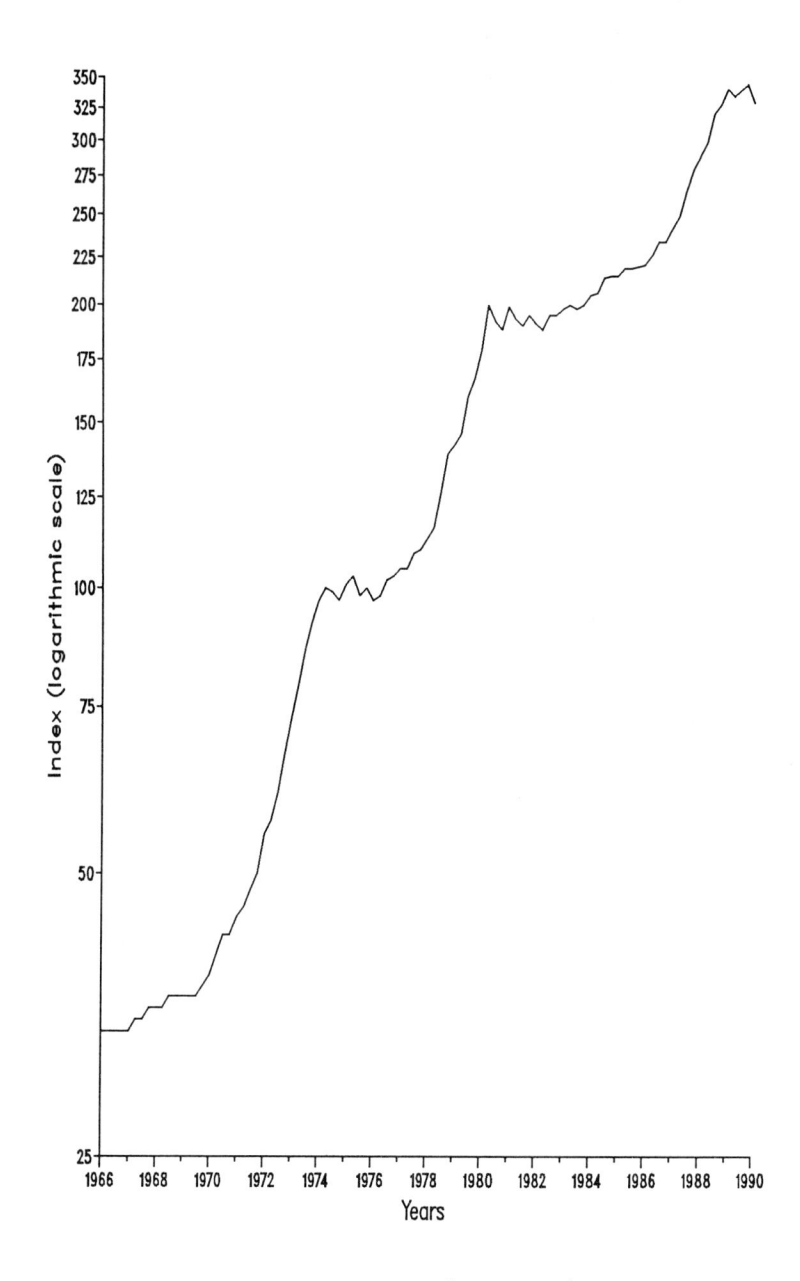

Figure 5.7.1 D L & E Tender Price Index (1976 = 100), 1966 Q1 – 1990 Q1

6

Cost Indices

6.1 BCIS BUILDING COST INDICES

Type of Index

Building cost index (BCI).

Series: Coverage and Breakdowns

Analyses of new building in the public and private sectors with breakdowns by building type as follows:

Series	*Table Reference*
BCIS General Building Cost Index (excluding M & E), 1971 to date	6.1.1
BCIS General Building Cost Index, 1971 to date	6.1.2
BCIS Steel Framed Construction Cost Index, 1971 to date	6.1.3
BCIS Concrete Framed Construction Cost Index, 1971 to date	6.1.4
BCIS Brick Construction Cost Index, 1971 to date	6.1.5
BCIS Mechanical and Electrical Engineering Cost Index, 1971 to date	6.1.6
Basic Labour Cost Index, 1974 to date	6.1.7
Basic Materials Cost Index, 1974 to date	6.1.8
Basic Plant Cost Index, 1974 to date	6.1.9

Base Dates and Period Covered

| 1974 Q1 | from 1971 or 1974 to date |
| 1985 | from 1984 to date |

The pre-1985 series are shown with 1985 = 100 in the tables below (converted arithmetically by the authors).

Frequency

Quarterly.

Geographical Coverage

United Kingdom.

Type and Source of Data

The inputs to the BCIS building cost indices are the Work Category Indices (Series 2) prepared by the Property Services Agency for use with the PSA Price Adjustment Formulae for Construction Contracts.

Method of Compilation

The BCIS building cost indices are based on cost models of average buildings. The following notes are based on those in the BCIS manual:
  'The BCIS Building Cost Indices consist of six indices, as follows:-

No. 11 - General Building Cost Index (excluding M & E).
No. 12 - General Building Cost index
No. 13 - Steel Framed Construction Cost Index
No. 14 - Concrete Framed Construction Cost Index
No. 15 - Brick Construction Cost Index
No. 16 - Mechanical and Electrical Engineering Cost Index

together with three input cost indices, as follows:-

No. 17 - Basic Labour Cost Index
No. 18 - Basic Materials Cost Index
No. 19 - Basic Plant Cost Index

The indices are based upon the PSA price adjustment formulae construction indices series 2 and as such take account of changes in the price of labour, materials and plant. The indices only measure the change in the cost of construction to the contractor and *do not* make any allowance for market conditions.
  The General Building Cost Index is compiled from the steel framed, concrete framed and brick construction cost indices in the proportion of 25%, 25% and 50% respectively.
  The General Building Cost Index (excluding M & E) is a similar index to the General Building Cost Index *but excludes* electrical, heating, ventilating and air conditioning and lift installations.
  The Mechanical and Electrical Engineering Cost Index includes solely those installations in the same ratio as used in the General Building Cost Index.
  All indices exclude any element of external works.

The Basic Labour Cost Index, the Basic Materials Cost Index and the Basic Plant Cost Index reflects the relevant element (Labour, Materials or Plant) only of the cost model of average buildings and includes all categories of work.

The indices were prepared by analysing forty bills of quantities into the building work and specialist engineering work categories. Because of the 25, 25, 50 per cent mix used in the General Index, ten steel framed, ten concrete framed and twenty brickwork construction jobs were chosen.

All measured items except for external works were allocated into the work groups. Preliminaries and provisional sums were not allocated, but as far as possible prime cost (pc) sums were put in the appropriate work group. If a pc sum involved a composite item, the item was split into the various work groups; e.g. with factory glazed metal windows, if the area of the window was known then the item would be allocated partly to glazing and partly to builders' general metalwork. If any appreciable amount of pc sums were found which could not be satisfactorily allocated, then the contract was not included.

Once the bill items had been allocated into work categories, the values were reduced to 1970 rates by using the PSA price adjustment formulae construction indices and converted to a percentage of the total. These percentages were then averaged for each type of construction to obtain a representative weighting for buildings of steel framed, concrete framed and brickwork construction.

The indices were derived by averaging the monthly PSA price adjustment formulae construction indices for each quarter and multiplying by the weighting for each work category. This gave an index based upon 1970 average = 100 which was then converted to the first quarter 1974 = 100. These have since been rebased to 1985 = 100.

In order to use the specialist engineering installation indices it was necessary to determine the labour and material proportions applicable to each index.

Structural Steelwork

The weighting for this index was determined from labour and material constants for steelwork from which the labour content for both the erection and fabrication of steelwork were formed and hence the relative proportions of labour and materials. The weighting used is 60 per cent labour and 40 per cent material.

Lift Work

The weighting used for the lift index is that defined in paragraph 59 of the JCT publication 'Standard Form of Building Contract Formula Rules' namely 47 per cent mechanical labour, 18 per cent mechanical materials and 12 per cent electrical materials.

For the electrical and heating, ventilating and air conditioning work a questionnaire was sent out to approximately 150 members asking for the labour/material weightings used by them on formula contracts. The replies received numbered 53, of which 45 were able to give figures. A total of 99 weightings for electrical contracts and 97 weightings for heating, ventilating and air conditioning contracts were obtained.

Electrical Work
A labour/material weighting of 40:60 was the ratio most commonly used (39 per cent), with a range of weightings from 18:82 to 60:40 being reported. The electrical labour/material weightings given by the sample were as follows:

39 per cent gave a weighting of 40:60 (labour/material)
12 per cent gave a weighting of 50:50
11 per cent gave a weighting of 45:55
10 per cent gave a weighting of 60:40
 7 per cent gave a weighting of 30:70
21 per cent remainder

Of the three most commonly occurring types of building (housing, schools and offices), the offices had the widest range of weightings, with a tendency towards a higher proportion of materials. Housing, however, had a higher labour content with an average weighting of 46:54 compared with the general average of 42:58.

The electrical weighting used in the BCIS indices is 40 per cent labour, 60 per cent material. This is the most commonly used in practice and is also the same as the average figure rounded off to the nearest 5 per cent.

Heating, Ventilating and Air Conditioning Work
With heating, ventilating and air conditioning work the most commonly occurring weighting was 35:65 labour:material (42 per cent of the sample) with a range of weightings from 13:87 to 60:40. Although the overall range of weightings was greater than for the electrical work, the majority of weightings were much closer together.

The heating, ventilating and air conditioning labour/material weightings given by the sample were as follows:

43 per cent gave a weighting of 35:65 (labour/materials)
23 per cent gave a weighting of 30:70
12 per cent gave a weighting of 40:60
 5 per cent gave a weighting of 20:80
17 per cent remainder

The heating, ventilating and air conditioning weighting used in the BCIS indices is 35 per cent labour, 65 per cent material and is similar to the electrical weighting in being the figure most commonly used in practice and is also the average figure rounded off. In addition to the foregoing the BCIS apply the monthly PSA price adjustment formulae construction indices to the weightings to produce a monthly index which is carried by the BCIS on-line service. Figures in the tables that follow are based on the PSA price adjustment formulae construction indices series 1 up to the fourth quarter 1977, from the first quarter 1978 they are based on the PSA price adjustment formulae construction indices series 2. The weightings for these indices are available on subscription from the BCIS.'

Commentary

The indices measure the changes in costs of labour, materials and plant (i.e. basic cost to the contractor) between one time period and another. They *do not* make any allowance for market conditions. Their main use, therefore, is in making an assessment of the change in cost to a contractor, for all or one particular factor, between one time period or another.

Publications

(a) Data Source

The BCIS *Quarterly Review of Building Prices* gives the General Building Cost Index and is available on subscription from Building Cost Information Service, 86/87 Clarence Street, Kingston-upon-Thames, Surrey, KT1 1RB. The remainder are available in the full subscription service and on-line service from the Building Cost Information Service as above.

(b) Description of Methodology

Details of the methodology are given in 'Cost study F7' available from the Building Cost Information Service as above.

Table 6.1.1 BCIS General Building Cost Index (Excluding M & E)

Base: 1985 = 100*

Year	Q1	Q2	Q3	Q4	Average
1971	17	18	18	18	18
1972	18	19	19	21	19
1973	22	22	23	24	23
1974	25	26	28	29	27
1975	31	33	35	36	34
1976	37	38	42	43	40
1977	44	46	48	48	47
1978	49	50	52	53	51
1979	54	56	61	63	59
1980	64	67	73	74	70
1981	74	76	78	80	77
1982	81	83	85	86	84
1983	86	88	91	91	89
1984	92	93	96	97	95
1985	98	99	102	102	100
1986	102	103	105	106	104
1987	106	108	110	111	110
1988	111	113	117	118	115
1989	119	122	126	127	124
1990	128**				
1991					
1992					
1993					
1994					
1995					

Source: BCIS.

*   Indices from 1971 to 1983 inclusive converted arithmetically by the authors from base of 1974 Q1 = 100 to 1985 = 100.
** Provisional.

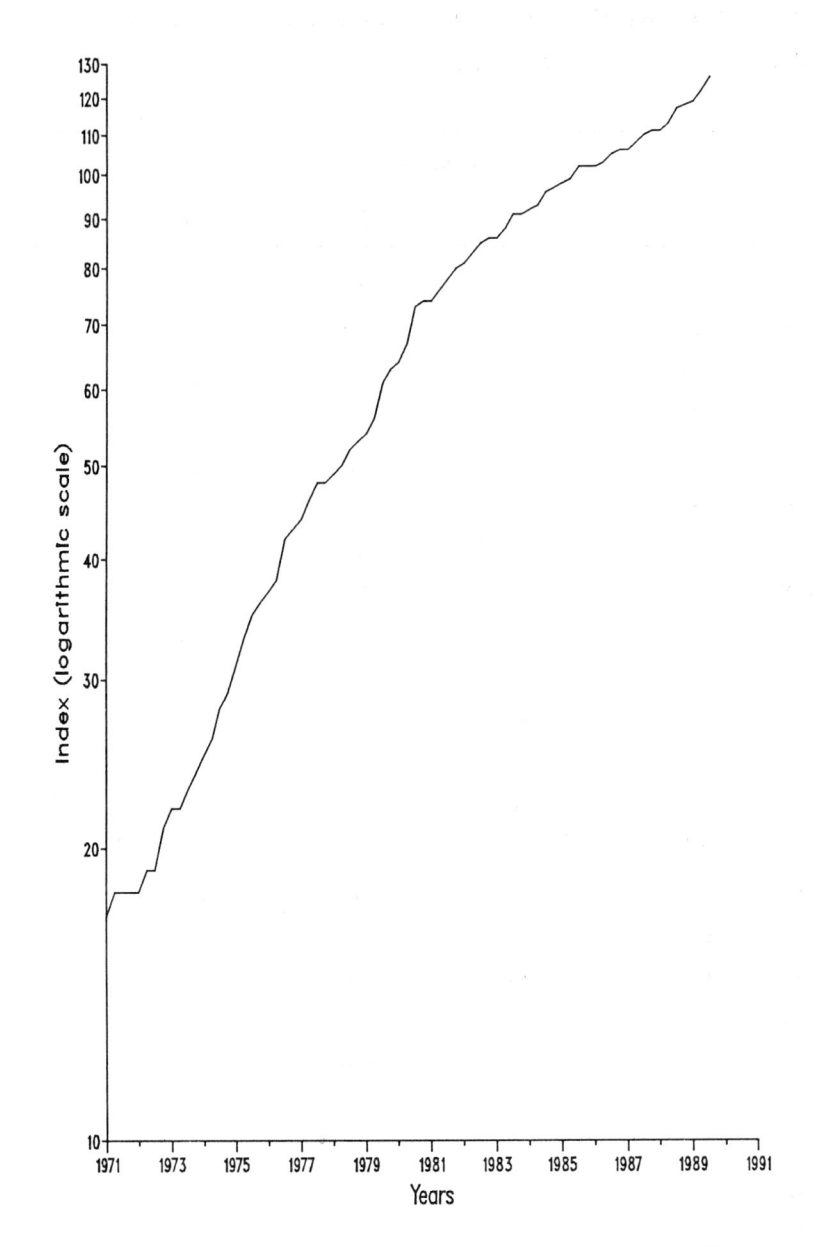

*Figure 6.1.1 BCIS General Building Cost Index (excluding M & E) –
(1985 = 100), 1971 Q1 – 1989 Q3*

Table 6.1.2 BCIS General Building Cost Index

Base: 1985 = 100*

Year	Q1	Q2	Q3	Q4	Average
1971	17	18	18	18	18
1972	18	19	19	21	19
1973	21	22	23	24	23
1974	25	27	28	29	27
1975	31	33	35	36	34
1976	37	38	41	42	40
1977	44	45	46	47	46
1978	49	49	51	53	50
1979	54	55	60	62	58
1980	65	67	72	73	69
1981	74	76	78	79	77
1982	81	83	85	85	84
1983	86	88	90	91	89
1984	92	94	96	96	95
1985	98	99	101	102	100
1986	102	103	105	106	104
1987	107	108	110	111	109
1988	112	114	117	118	115
1989	120	122	126	127	124
1990	128**				
1991					
1992					
1993					
1994					
1995					

Source: BCIS.

* Indices from 1971 to 1983 inclusive converted arithmetically by the authors from base of 1974 Q1 = 100 to 1985 = 100.

** Provisional.

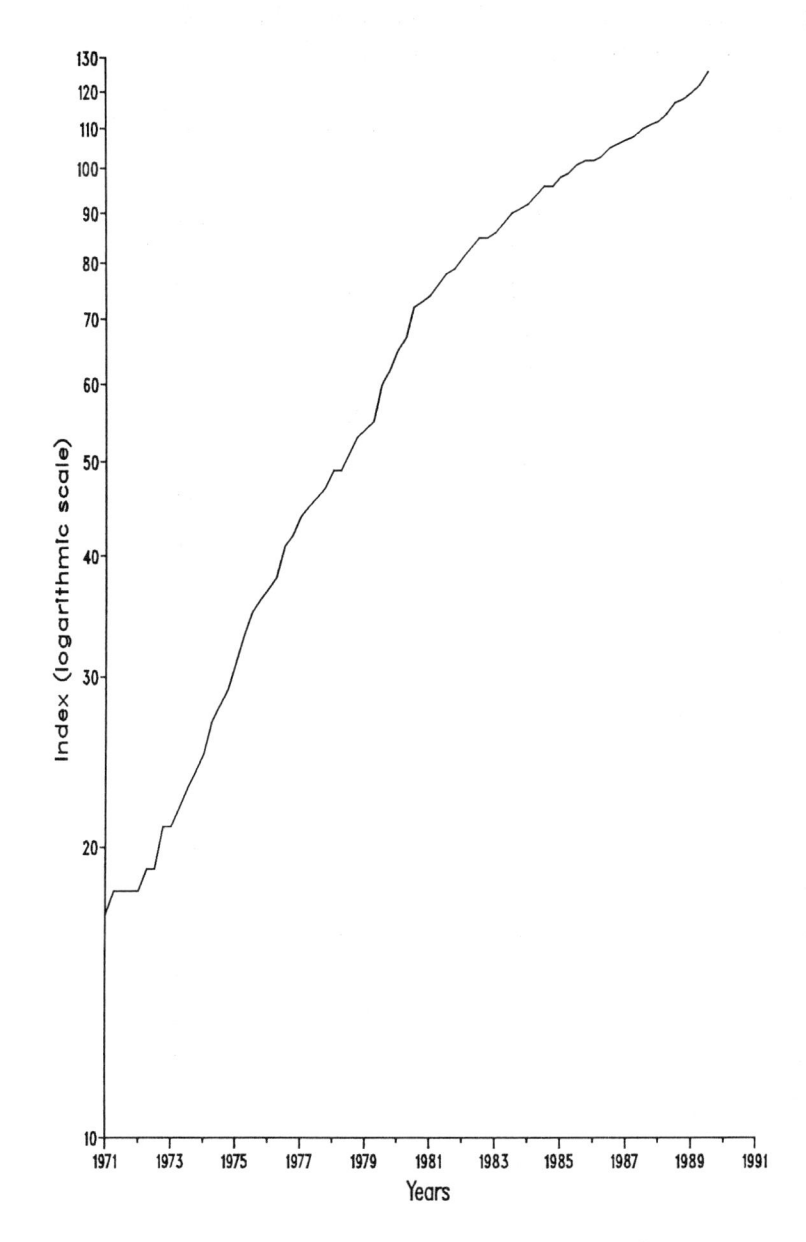

Figure 6.1.2 BCIS General Building Cost Index (1985 = 100), 1971 Q1 – 1989 Q3

Table 6.1.3 BCIS Steel Framed Construction Cost Index

Base: 1985 = 100*

Year	Q1	Q2	Q3	Q4	Average
1971	17	18	18	18	18
1972	18	19	19	21	19
1973	21	21	22	23	22
1974	25	27	28	29	27
1975	31	33	35	36	34
1976	37	38	41	42	40
1977	44	45	47	48	46
1978	49	50	52	53	51
1979	54	56	60	63	58
1980	65	67	72	73	69
1981	74	76	77	79	77
1982	81	83	85	85	84
1983	85	88	90	91	89
1984	92	93	96	96	94
1985	98	99	101	102	100
1986	102	103	105	106	104
1987	107	108	111	111	109
1988	112	114	117	118	115
1989	120	123	126	127	124
1990	128**				
1991					
1992					
1993					
1994					
1995					

Source: BCIS.

* Indices from 1971 to 1983 inclusive converted arithmetically by the authors from base of 1974 Q1 = 100 to 1985 = 100.

** Provisional.

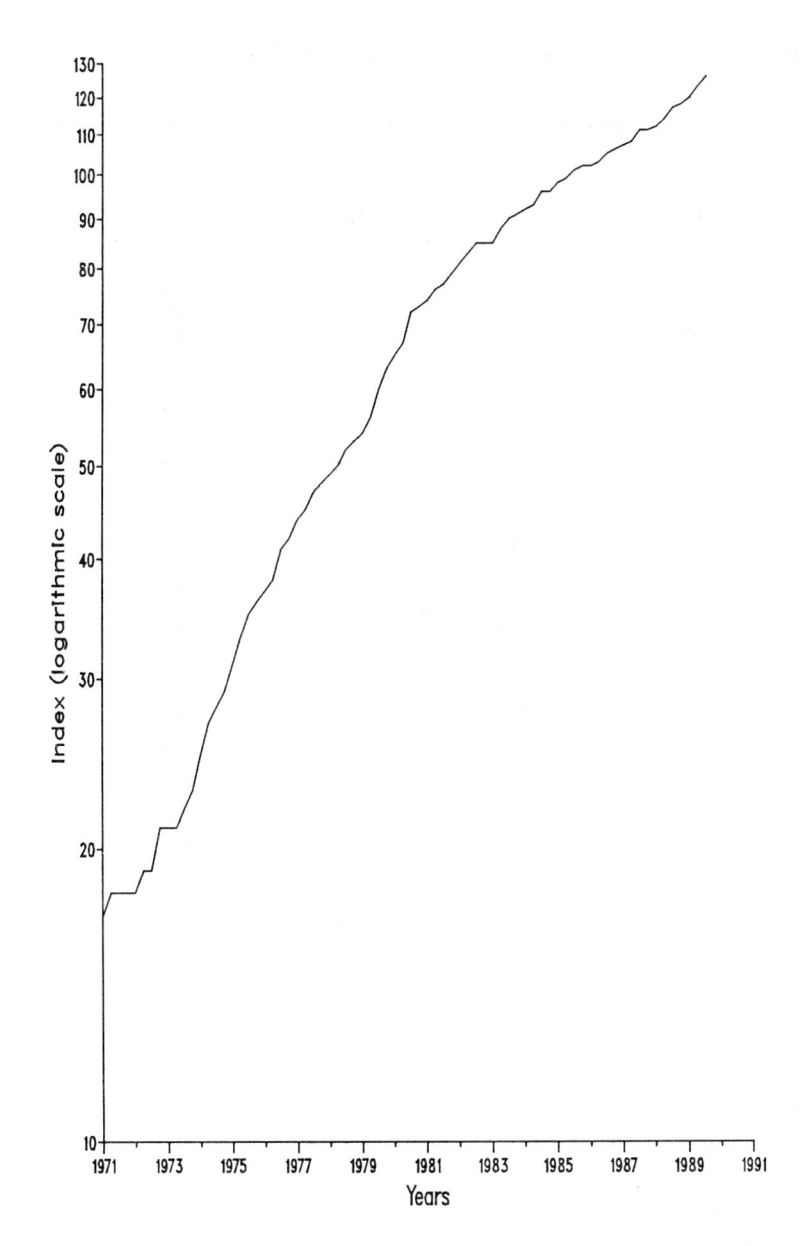

Figure 6.1.3 BCIS Steel Framed Construction Cost Index (1985 = 100)
1971 Q1 – 1989 Q3

Table 6.1.4 BCIS Concrete Framed Construction Cost Index

Base: 1985 = 100*

Year	Q1	Q2	Q3	Q4	Average
1971	17	18	18	18	18
1972	19	19	19	21	20
1973	21	22	23	24	23
1974	25	27	28	29	27
1975	31	33	35	36	34
1976	36	38	41	42	39
1977	43	45	46	47	45
1978	48	49	51	52	50
1979	54	55	60	63	58
1980	65	67	72	73	69
1981	74	76	78	79	77
1982	81	83	85	86	84
1983	86	88	91	91	89
1984	92	94	96	96	95
1985	98	99	101	102	100
1986	102	103	105	105	104
1987	106	107	110	110	108
1988	111	113	116	118	115
1989	120	122	126	127	124
1990	128**				
1991					
1992					
1993					
1994					
1995					

Source: BCIS.

* Indices from 1971 to 1983 inclusive converted arithmetically by the authors from base of 1974 Q1 = 100 to 1985 = 100.

** Provisional.

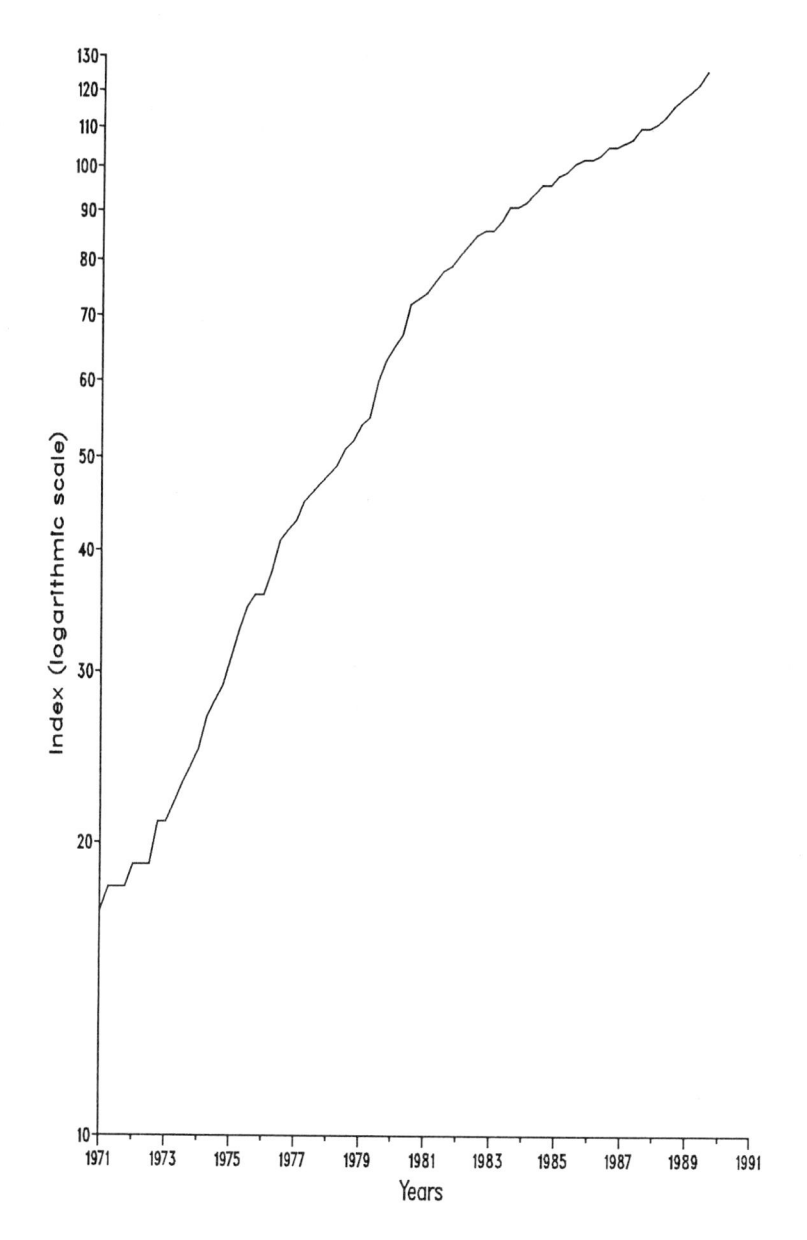

*Figure 6.1.4 BCIS Concrete Framed Construction Cost Index (1985 = 100)
1971 Q1 – 1989 Q3*

Table 6.1.5 BCIS Brick Construction Cost Index

Base: 1985 = 100*

Year	Q1	Q2	Q3	Q4	Average
1971	17	18	18	18	18
1972	18	19	19	21	19
1973	21	22	23	24	23
1974	25	27	28	29	27
1975	31	33	35	35	34
1976	37	38	41	42	40
1977	44	45	47	47	46
1978	48	49	51	52	50
1979	54	55	60	62	58
1980	65	67	72	73	69
1981	74	76	78	79	77
1982	81	83	85	86	84
1983	86	88	91	92	89
1984	92	94	96	97	95
1985	98	99	101	102	100
1986	102	103	105	106	104
1987	107	108	111	111	109
1988	112	114	117	118	115
1989	120	122	126	127	124
1990	128**				
1991					
1992					
1993					
1994					
1995					

Source: BCIS.

*   Indices from 1971 to 1983 inclusive converted arithmetically by the authors from base of 1974 Q1 = 100 to 1985 = 100.
** Provisional.

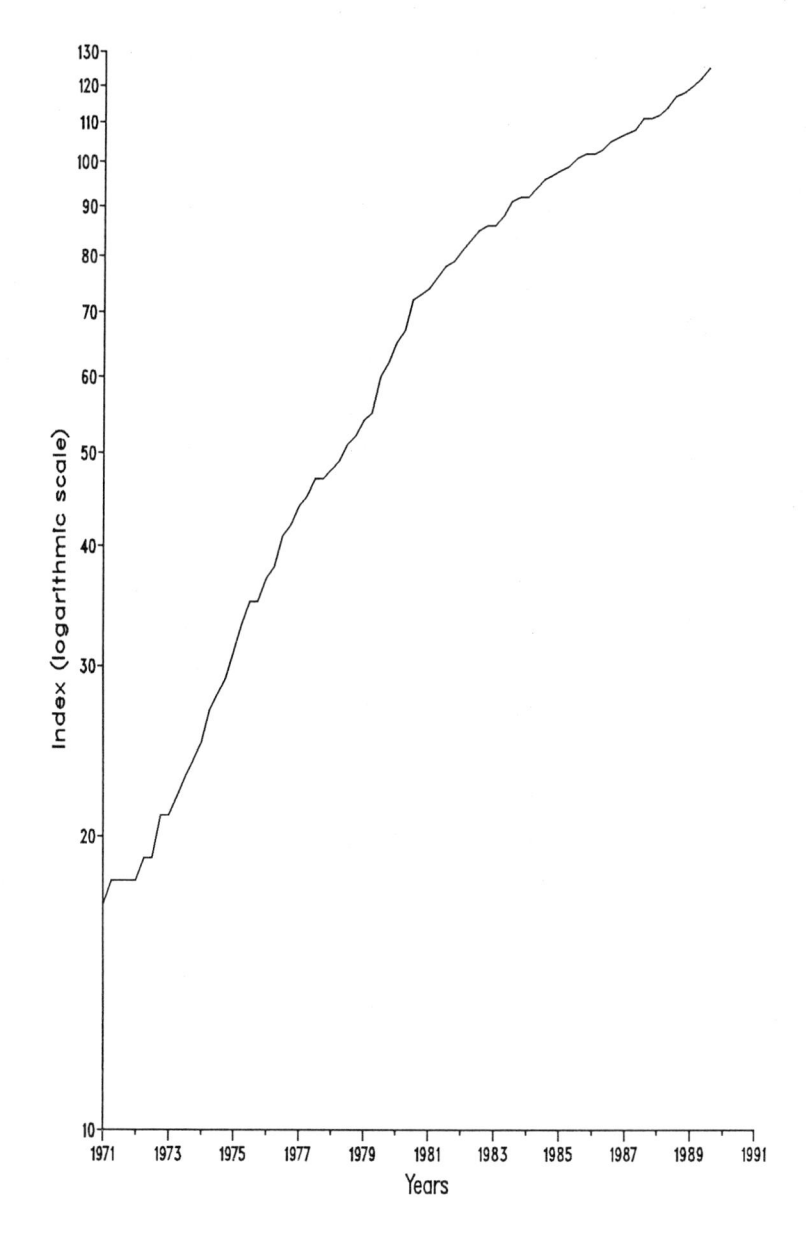

Figure 6.1.5 BCIS Brick Construction Cost Index (1985 = 100), 1971 Q1 – 1989 Q3

Table 6.1.6 BCIS Mechanical and Electrical Engineering Cost Index

Base: 1985 = 100*

Year	Q1	Q2	Q3	Q4	Average
1971	17	18	18	18	18
1972	19	19	19	20	19
1973	20	21	22	23	22
1974	25	28	28	29	28
1975	31	33	33	35	33
1976	35	37	38	40	38
1977	41	42	43	44	43
1978	47	48	49	51	49
1979	53	55	58	61	57
1980	66	68	70	72	69
1981	75	76	77	78	77
1982	81	84	84	85	84
1983	86	89	91	92	90
1984	93	94	96	96	95
1985	98	100	101	102	100
1986	103	104	105	106	105
1987	108	110	111	112	110
1988	114	116	117	119	117
1989	121	124	125	127	124
1990	129**				
1991					
1992					
1993					
1994					
1995					

Source: BCIS.

* Indices from 1971 to 1983 inclusive converted arithmetically by the authors from base of 1974 Q1 = 100 to 1985 = 100.

** Provisional.

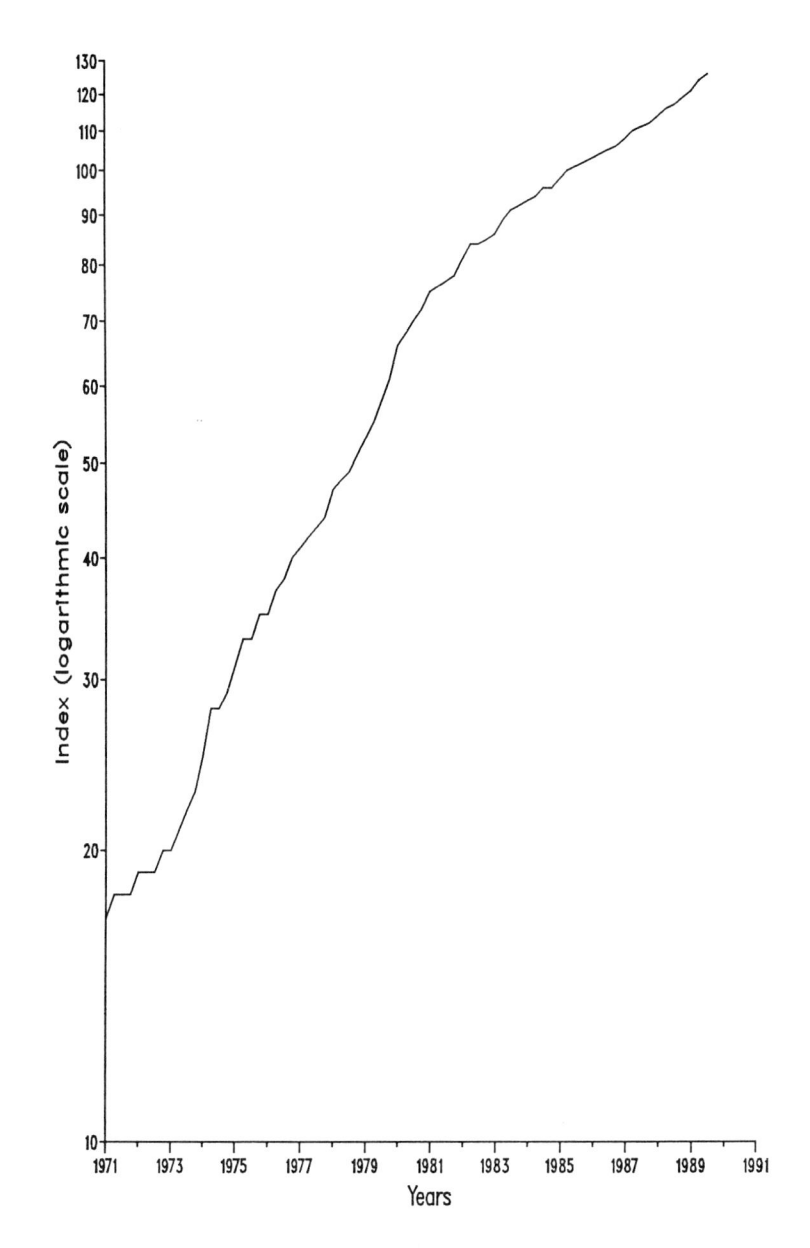

Figure 6.1.6 BCIS Mechanical & Electrical Engineering Cost Index (1985 = 100), 1971 Q1 – 1989 Q3

Table 6.1.7 BCIS Basic Labour Cost Index

Base: 1985 = 100*

Year	Q1	Q2	Q3	Q4	Average
1974	24	24	27	28	26
1975	30	32	36	36	34
1976	37	37	41	41	39
1977	41	42	44	44	43
1978	45	46	49	50	48
1979	52	52	59	61	56
1980	62	63	72	74	68
1981	75	76	78	81	78
1982	83	84	87	87	85
1983	87	89	93	93	91
1984	93	94	98	97	96
1985	97	98	102	102	100
1986	103	104	108	108	106
1987	109	110	113	114	112
1988	114	115	120	121	118
1989	122	123	129	130	126
1990	131**				
1991					
1992					
1993					
1994					
1995					

Source: BCIS.

* Indices from 1974 to 1983 inclusive converted arithmetically by the authors from base of 1974 Q1 = 100 to 1985 = 100.

** Provisional.

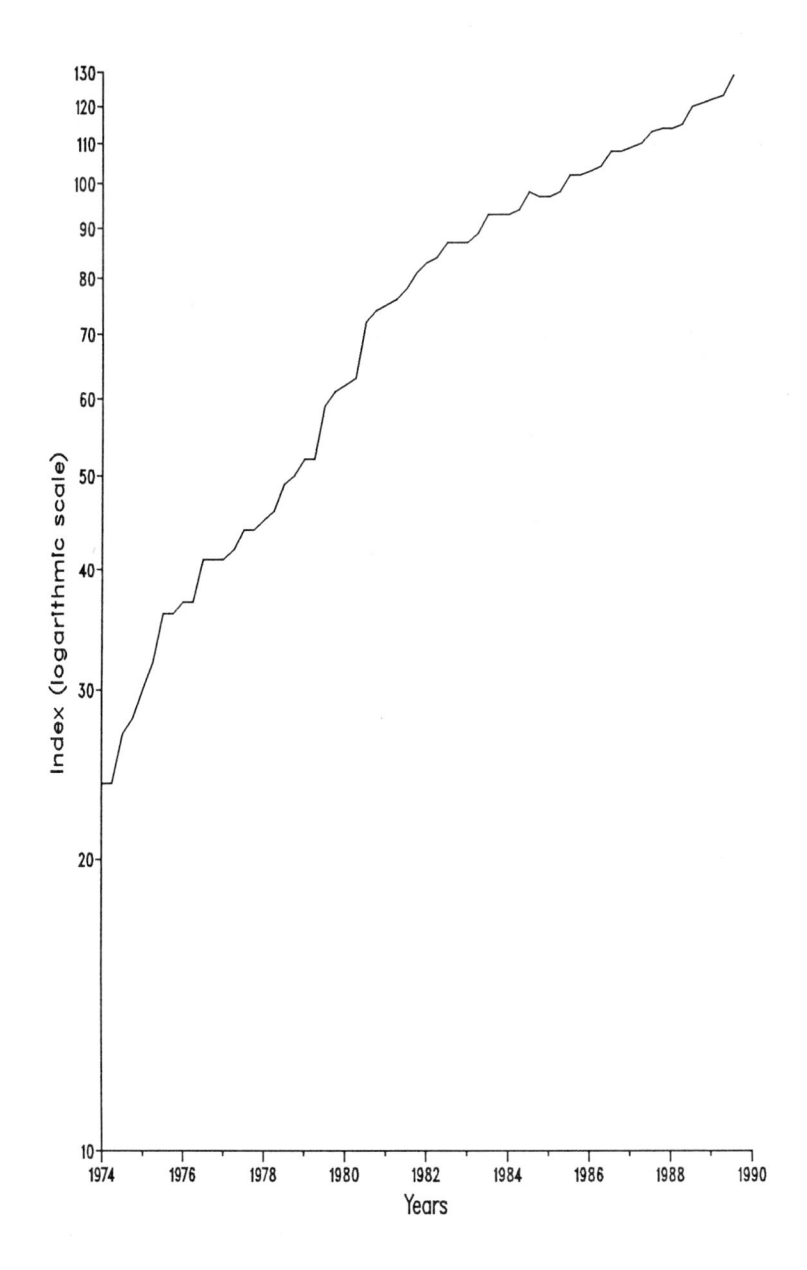

Figure 6.1.7 BCIS Basic Labour Cost Index (1985 = 100), 1974 Q1 – 1989 Q3

Table 6.1.8 BCIS Basic Materials Cost Index

Base: 1985 = 100*

Year	Q1	Q2	Q3	Q4	Average
1974	26	27	29	29	28
1975	31	32	33	35	33
1976	37	39	42	44	41
1977	45	47	49	50	48
1978	50	52	53	54	52
1979	56	58	61	64	60
1980	67	71	72	73	71
1981	74	76	77	78	76
1982	80	82	83	84	82
1983	85	88	89	91	88
1984	92	94	95	96	94
1985	98	100	101	101	100
1986	102	103	103	104	103
1987	105	107	108	109	107
1988	110	112	114	117	113
1989	118	122	123	124	122
1990	126**				
1991					
1992					
1993					
1994					
1995					

Source: BCIS.

* Indices from 1974 to 1983 inclusive converted arithmetically by the authors from base of 1974 Q1 = 100 to 1985 = 100.

** Provisional.

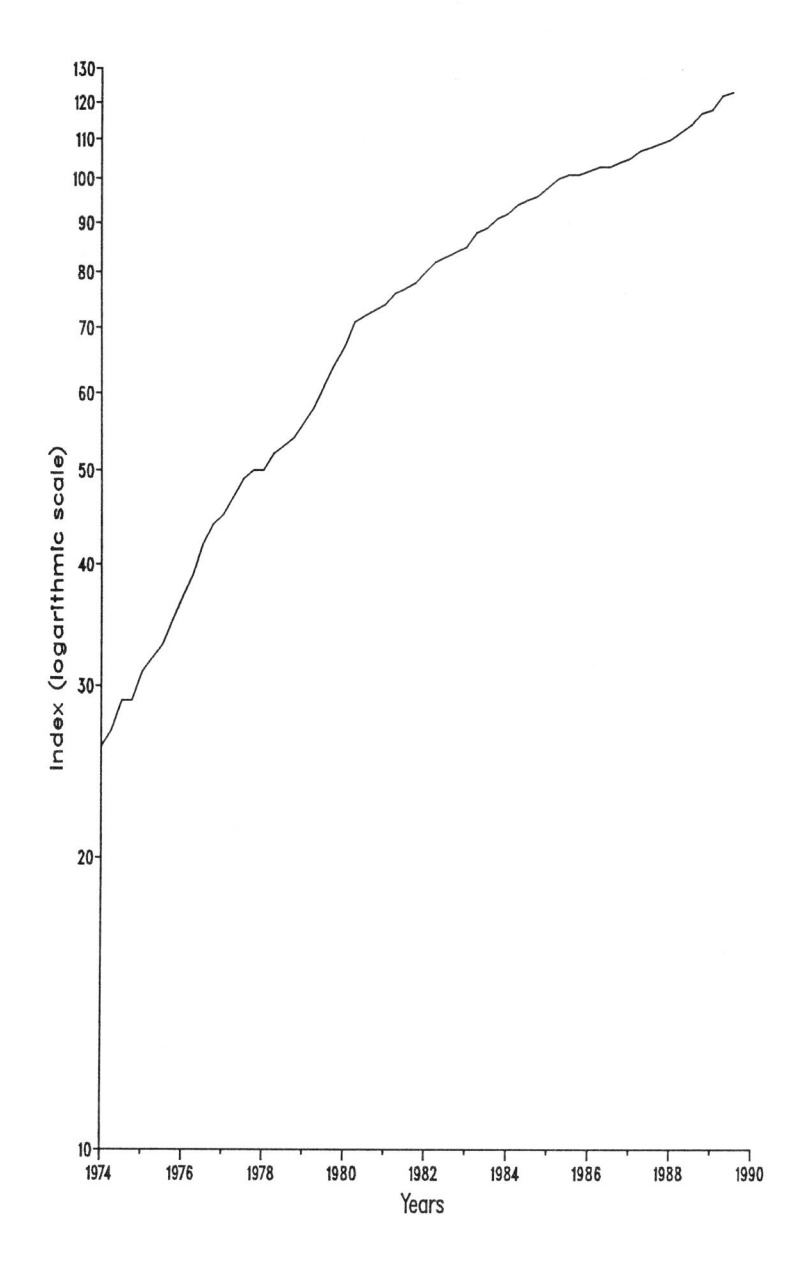

Figure 6.1.8 BCIS Basic Materials Cost Index (1985 = 100), 1974 Q1 – 1989 Q3

Table 6.1.9 BCIS Basic Plant Cost Index

Base: 1985 = 100*

Year	Q1	Q2	Q3	Q4	Average
1974	24	25	27	28	26
1975	30	32	34	36	33
1976	37	38	40	42	39
1977	43	45	47	48	46
1978	48	49	51	52	50
1979	53	54	60	61	57
1980	64	65	72	72	68
1981	73	74	77	81	76
1982	81	82	85	86	84
1983	87	88	91	89	89
1984	90	91	94	96	93
1985	99	98	102	101	100
1986	100	98	98	100	99
1987	101	102	106	106	104
1988	107	107	110	110	109
1989	111	113	117	119	115
1990	121**				
1991					
1992					
1993					
1994					
1995					

Source: BCIS.

* Indices from 1974 to 1983 inclusive converted arithmetically by the authors from base of 1974 Q1 = 100 to 1985 = 100.

** Provisional.

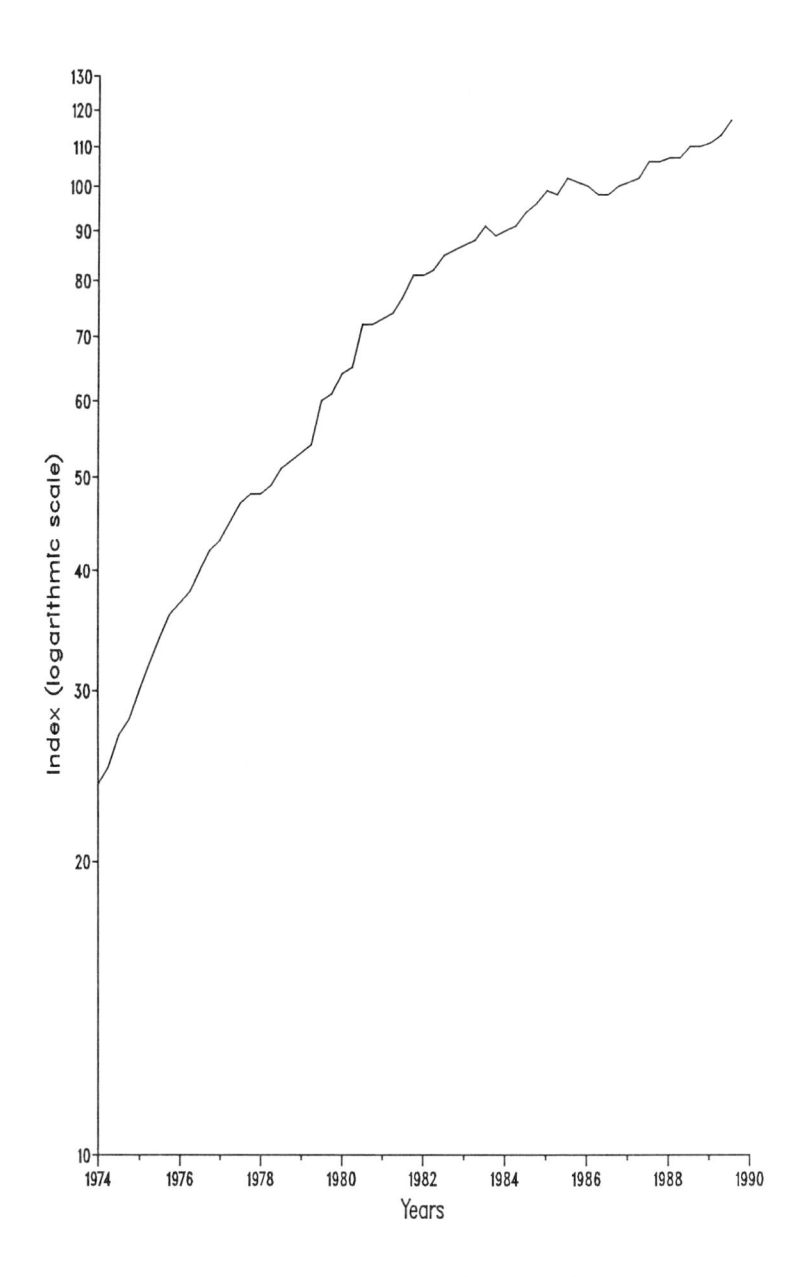

Figure 6.1.9 BCIS Basic Plant Cost Index (1985 = 100), 1974 Q1 – 1989 Q3

6.2 SPON'S COST INDICES

Type of Index

Building cost index (BCI).

Series: Coverage and Breakdowns

Analyses of new building in the public and private sectors with breakdowns by building type and content as follows:

Series	Table Reference
Spon's Building Cost Index, 1966 to date	6.2.1
Spon's Mechanical Services Cost Index, 1966 to date	6.2.2
Spon's Electrical Services Cost Index, 1966 to date	6.2.3
Spon's Constructed Civil Engineering Cost Index, 1971 to date	6.2.4
Spon's Constructed Landscaping (Hard Surfacing and Planting) Cost Index, 1976 to date	6.2.5

Base Dates and Period Covered

1956 January	from 1956 to 1975
1965 January	from 1965 to 1981
1970	from 1970 to 1980
1976	from 1970 to date

The pre-1976 series are shown with 1976 = 100 in the tables below (converted arithmetically by the authors).

Frequency

Quarterly

Geographical Coverage

United Kingdom

Type and Source of Data

The inputs to these indices are the relevant labour rates based on either a wage sheet or a notional gang and the changes in cost of materials are based on the relevant indices prepared by the Department of Trade and Industry.

Method of Compilation

The description of the Spon's Building Costs Index which appeared in the latest edition of the *Spon's Architects' and Builders' Price Book* is as follows:

'This table (6.2.1) reflects the fluctuations...in wages and materials costs to the builder. In compiling the table the proportion of labour to material has been assumed to be 40:60. The wages element has been assessed from a contract wages sheet revalued for each variation in labour costs while the changes in the cost of materials have been based on the indices prepared by the Department of Trade and Industry. No allowance has been made for changes in productivity, plus rates or hours worked which may occur in particular conditions and localities.'

The description of the Spon's Mechanical Services Cost Index and the Spon's Electrical Services Cost Index which appeared in the latest edition of the *Spon's Mechanical and Electrical Services Price Book* is as follows:

'The following tables (6.2.2 and 6.2.3) reflect the major changes in cost to contractors but do not necessarily reflect changes in tender levels. In addition to changes in labour and materials costs tenders are affected by other factors such as the degree of competition in the particular industry and area where the work is to be carried out, the availability of labour and general economic situation. This has meant in recent years that when there has been an abundance of work tender levels have often increased at a greater rate than can be accounted for by increases in basic labour and material costs and, conversely, when there is a shortage of work this has often resulted in keener tenders. Allowances for these factors are impossible to assess on a general basis and can only be based on experience and knowledge of the particular circumstances. In compiling the tables the cost of labour has been calculated on the basis of a notional gang as set out elsewhere in the book. The proportion of labour to materials has been assumed as follows: Mechanical Services 30:70, Electrical Services 50:50.'

The description of the Spon's Constructed Civil Engineering Cost Index which appeared in the latest edition of *Spon's Civil Engineering and Highway Works Price Book* is as follows:

'It is important to distinguish between costs and tender prices; civil engineering costs are the costs incurred by a contractor in the course of his business; civil engineering tender prices ar the prices for which a contractor undertakes work. Tender prices will be based on contractor's costs but will also take into account market considerations such as the availability of labour and materials and the prevailing workload for civil engineering contractors. This can mean that in a period when work is scarce tender prices may fall as costs are rising while when there is plenty of work prices will tend to increase at a faster rate than costs.

Cost indices for labour, plant and materials in civil engineering work are compiled and maintained by the Property Services Agency as Technical Secretariat to the working group (for NEDO price adjustment formula indices). They are published in a monthly bulletin by HMSO and are reproduced here with their permission (but not included in this book). These indices were formerly known as NEDO or "Baxter" indices. They comprise 14 indices derived from government sources. Two of them, DERV fuel (index 8) and light re-rolled bars and sections (index 11b), are little used and a

158   *Spon's handbook of construction cost and price indices*

further two, indices 12 and 13, represent materials and labour specifically for structural steelwork. Thus there are 10 current indices for general civil engineering work; the reference numbers and titles are listed below.

Index No.

1 Cost of labour
2 Cost of providing and maintaining constructional plant
3 Sand and gravel (ex pit or works)
4 Engineering bricks delivered
5 Cement delivered
6 Cast and spun iron pipes and fittings
7 Coated limestone roadstone
9 Gas oil fuel
10 Imported softwood
11a Steel for reinforcement (cut, bent and delivered)

Although the above indices are prepared and published in order to provide a common basis for calculating reimbursement of increased costs during the course of a contract, they also present time series of cost indices for the main components of civil engineering work. They can therefore be used as the basis of an index for civil engineering work. The method used here is to construct a composite index by allocating weightings to each of the 10 indices, the weightings being established from an analysis of actual projects. The composite index is calculated by applying these weightings to the appropriate price adjustment formula indices and totalling the results; this index is...presented with a base of 1970.'

The description of the Spon's Constructed Landscaping (Hard Surfacing and Planting) Cost Index which appeared in the latest edition of *Spon's Landscape and External Works Price Book* is as follows:

'The purpose of this section is to show changes in the cost of carrying out landscaping work (hard surfacing and planting) since 1976. It is important to distinguish between costs and tender prices: the following table (6.2.5) reflects the change in cost to contractors but does not necessarily reflect changes in tender prices. In addition to changes in labour and material costs, which are reflected in the indices given below, tender prices are also affected by factors such as the degree of competition at the time of tender and in the particular area where the work is to be carried out; the availability of labour and materials and the general economic situation. This can mean that in a period when work is scarce tender prices may fall despite the fact that costs are rising and when there is plenty of work available, tender prices may increase at a faster rate than costs.

A Constructed Cost Index based on PSA *Price Adjustment Formulae for Construction Contracts: Series 2*. Cost indices for the various trades employed in a building contract are published monthly by HMSO and are reproduced in the technical press. The indices comprise 49 Building Work indices plus seven "Appendices" and other specialist indices. The Building Work indices are compiled by monitoring the cost of labour and materials for each category and applying a weighting to these to calculate a single index.

Cost indices 159

Although the PSA indices are prepared for use with price adjustment formulae for calculating reimbursement of increased costs during the course of a contract, they also present a time series of cost indices for the main components of landscaping projects. They can therefore be used as the basis of an index for landscaping purposes.

The method used here is to construct a composite index by allocating weightings to the indices representing the usual work categories found in a landscaping project, the weightings being established from an analysis of actual projects. These weightings total 100 in 1976 and the composite index is calculated by applying the appropriate weightings to the appropriate PSA indices on a monthly basis, which is then compiled into a quarterly index and rebased to 1976 = 100. The Constructed Landscaping (Hard Surfacing and Planting) Cost Index is based on approximately 50 per cent of soft landscaping area and 50 per cent hard external works.'

Commentary

The indices measure the changes in costs of labour, materials and plant (i.e. basic cost to the contractor) between one time period and another. They *do not* make any allowance for market conditions. Their main use, therefore, is in making an assessment of the change in cost to a contractor, for all or one particular factor, between one time period or another.

Publications

(a) Data Source

Spon's Architects' and Builders' Price Book
Spon's Mechanical and Electrical Services Price Book
Spon's Civil Engineering and Highway Works Price Book
Spon's Landscape and External Works Price Book
Spon's Price Book Update (three times a year, available free of charge to all purchasers of Spon's Price Books.

(b) Description of Methodology

Full details of the relevant methodology are given in each of the *Price Books* mentioned above available from E & F N Spon, 2-6 Boundary Row, London SE1 8HN.

Table 6.2.1 Spon's Building Costs Index

Base: 1976 = 100*

Year	Q1	Q2	Q3	Q4	Average
1966	31	32	33	33	33
1967	33	34	34	35	34
1968	35	36	36	37	36
1969	37	37	38	38	38
1970	40	41	42	43	41
1971	43	45	46	46	45
1972	46	47	49	55	49
1973	56	56	59	61	58
1974	64	67	71	73	69
1975	78	82	89	90	85
1976	93	97	104	107	100
1977	109	112	116	117	114
1978	118	120	127	129	124
1979	131	135	149	153	142
1980	157	161	180	181	170
1981	182	185	195	199	190
1982	203	206	214	216	210
1983	217	219	227	229	223
1984	230	232	239	241	236
1985	243	245	252	254	249
1986	256	258	266	267	262
1987	270	272	281	282	276
1988	284	286	299	302	293
1989	305	307	322	323	314
1990	325**				
1991					
1992					
1993					
1994					
1995					

Source: *Spon's Architects' and Builders' Price Book* and *Building*

* Indices from 1966 to 1975 inclusive converted arithmetically by the authors from base of 1970 = 100 to 1976 = 100.
** Provisional.

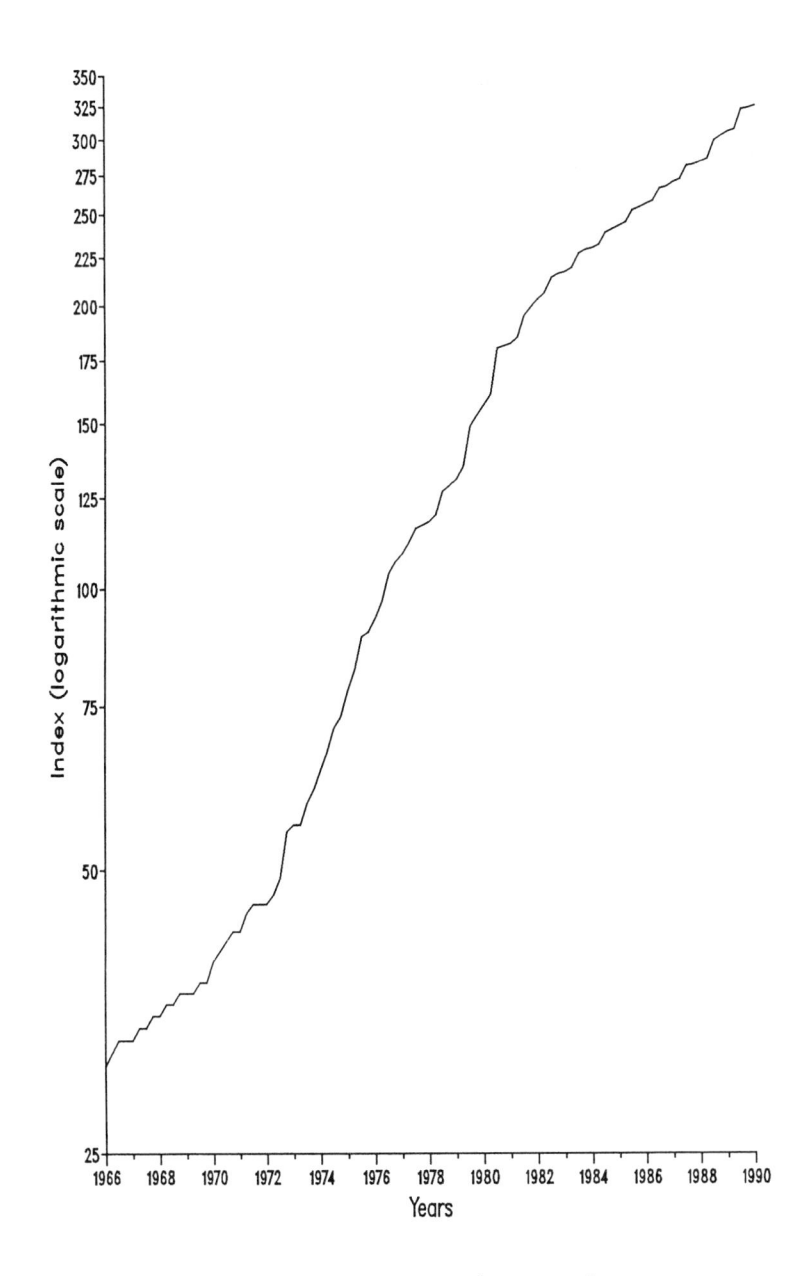

Figure 6.2.1. Spon's Building Costs Index (1976 = 100), 1966 Q1 – 1990 Q1

Table 6.2.2 Spon's Mechanical Services Cost Index

Base: 1976 = 100*

Year	Q1	Q2	Q3	Q4	Average
1966	33	33	34	34	33
1967	35	35	36	36	35
1968	37	37	38	38	38
1969	39	39	39	39	39
1970	41	42	42	43	42
1971	44	45	45	46	45
1972	49	50	50	51	50
1973	52	53	55	59	55
1974	64	70	73	76	71
1975	80	83	87	90	85
1976	92	97	103	107	100
1977	110	112	113	115	113
1978	121	123	126	133	126
1979	138	141	150	156	146
1980	165	171	172	174	171
1981	181	185	187	190	186
1982	197	201	201	201	200
1983	204	205	208	211	207
1984	215	221	223	226	221
1985	232	236	236	236	235
1986	238	240	239	243	240
1987	244	249	251	254	250
1988	257	265	268	272	266
1989	276	284	284	286	283
1990	287**				
1991					
1992					
1993					
1994					
1995					

Source: *Spon's Mechanical and Electrical Services Price Book* and *Building*.

* Indices from 1966 to 1969 inclusive converted arithmetically by the authors from base of 1970 = 100 to 1976 = 100.

** Provisional.

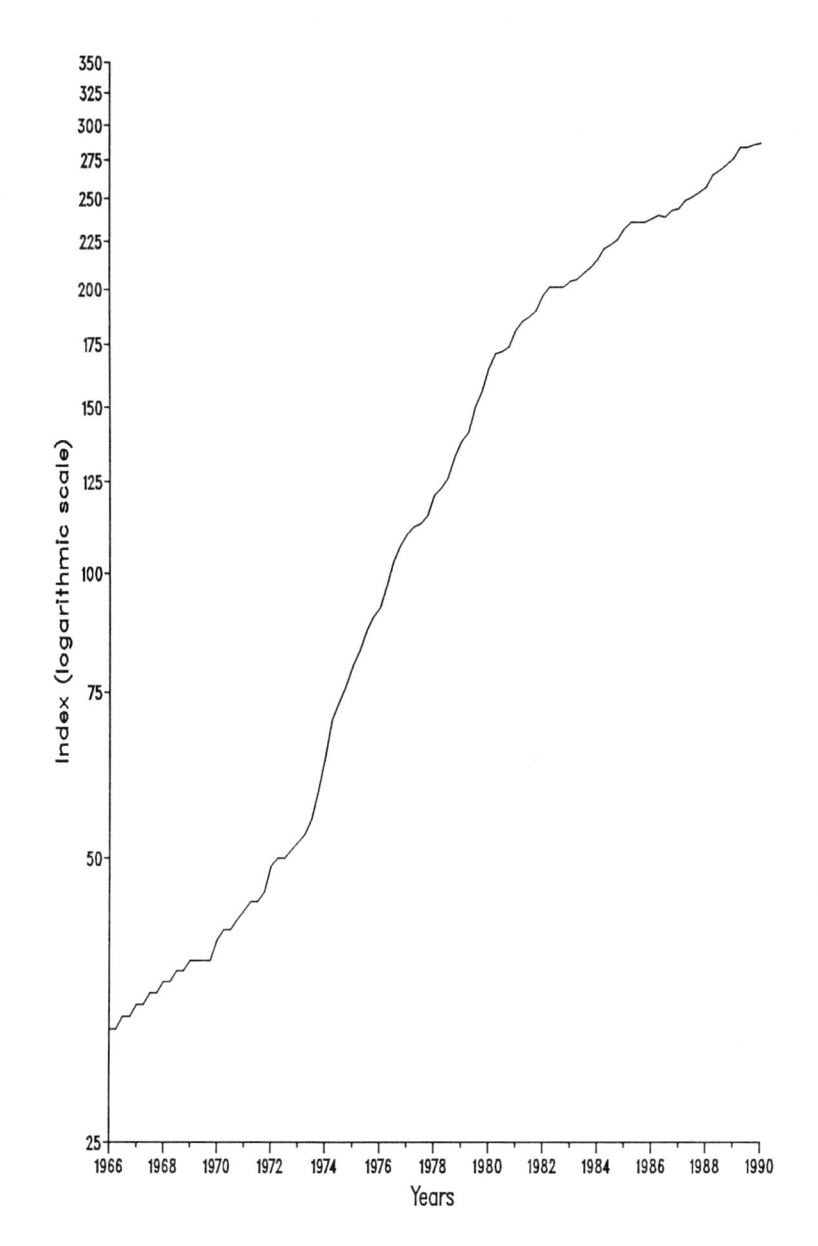

Figure 6.2.2 Spon's Mechanical Services Cost Index (1976 = 100), 1966 Q1 – 1990 Q1

Table 6.2.3 Spon's Electrical Services Cost Index

Base: 1976 = 100*

Year	Q1	Q2	Q3	Q4	Average
1966	28	30	30	31	30
1967	32	33	33	34	33
1968	36	36	37	37	36
1969	37	37	38	40	38
1970	40	41	41	43	41
1971	45	45	45	45	45
1972	48	49	50	51	50
1973	51	54	56	58	55
1974	64	67	70	71	68
1975	80	81	86	88	84
1976	95	100	102	103	100
1977	109	110	110	111	110
1978	121	128	129	132	128
1979	143	145	147	154	147
1980	169	169	177	192	177
1981	195	197	199	200	198
1982	202	211	211	212	209
1983	214	225	225	226	223
1984	228	229	236	237	233
1985	240	240	247	250	244
1986	251	249	249	255	251
1987	268	269	271	274	271
1988	286	289	290	293	290
1989	306	306	306	308	307
1990	322**				
1991					
1992					
1993					
1994					
1995					

Source: *Spon's Mechanical and Electrical Services Price Book* and *Building*

*   Indices from 1966 to 1969 inclusive converted arithmetically by the authors from base of 1970 = 100 to 1976 = 100.
** Provisional.

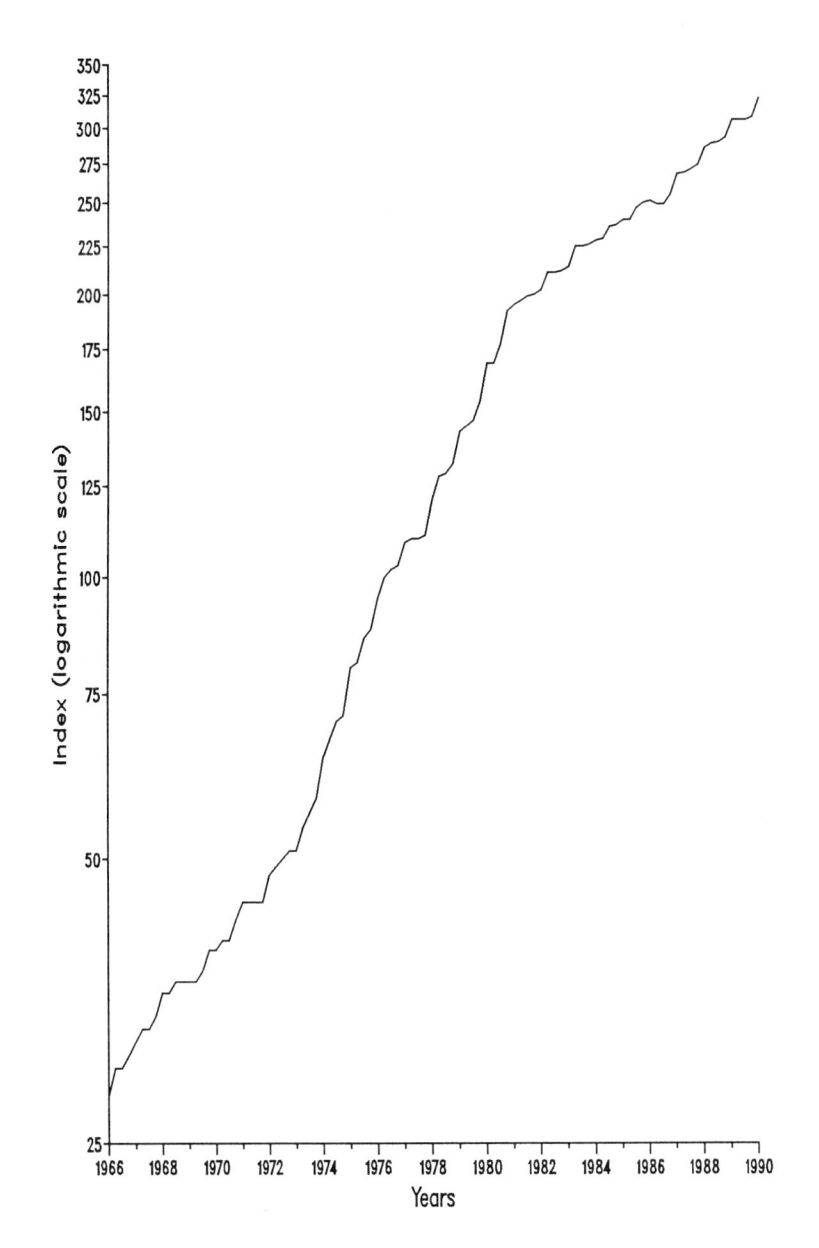

Figure 6.2.3 Spon's Electrical Services Cost Index (1976 = 100), 1966 Q1 – 1990 Q1

Table 6.2.4 Spon's Constructed Civil Engineering Cost Index

Base: 1970 = 100

Year	Q1	Q2	Q3	Q4	Average
1971	106	111	113	113	111
1972	114	116	119	129	119
1973	130	132	137	142	135
1974	154	166	175	181	169
1975	198	209	220	227	213
1976	241	250	269	279	260
1977	291	303	314	319	307
1978	322	330	341	348	335
1979	357	373	409	420	390
1980	440	462	494	498	474
1981	509	524	541	558	533
1982	572	578	594	608	588
1983	613	625	640	639	629
1984	649	663	676	671	665
1985	690	698	712	710	703
1986	698	689	685	693	691
1987	701	709	728	730	717
1988	736	743	760	763	751
1989	767	786	809	825**	797**
1990					
1991					
1992					
1993					
1994					
1995					

Source: *Spon's Civil Engineering and Highway Works Price Book*

**   Provisional.

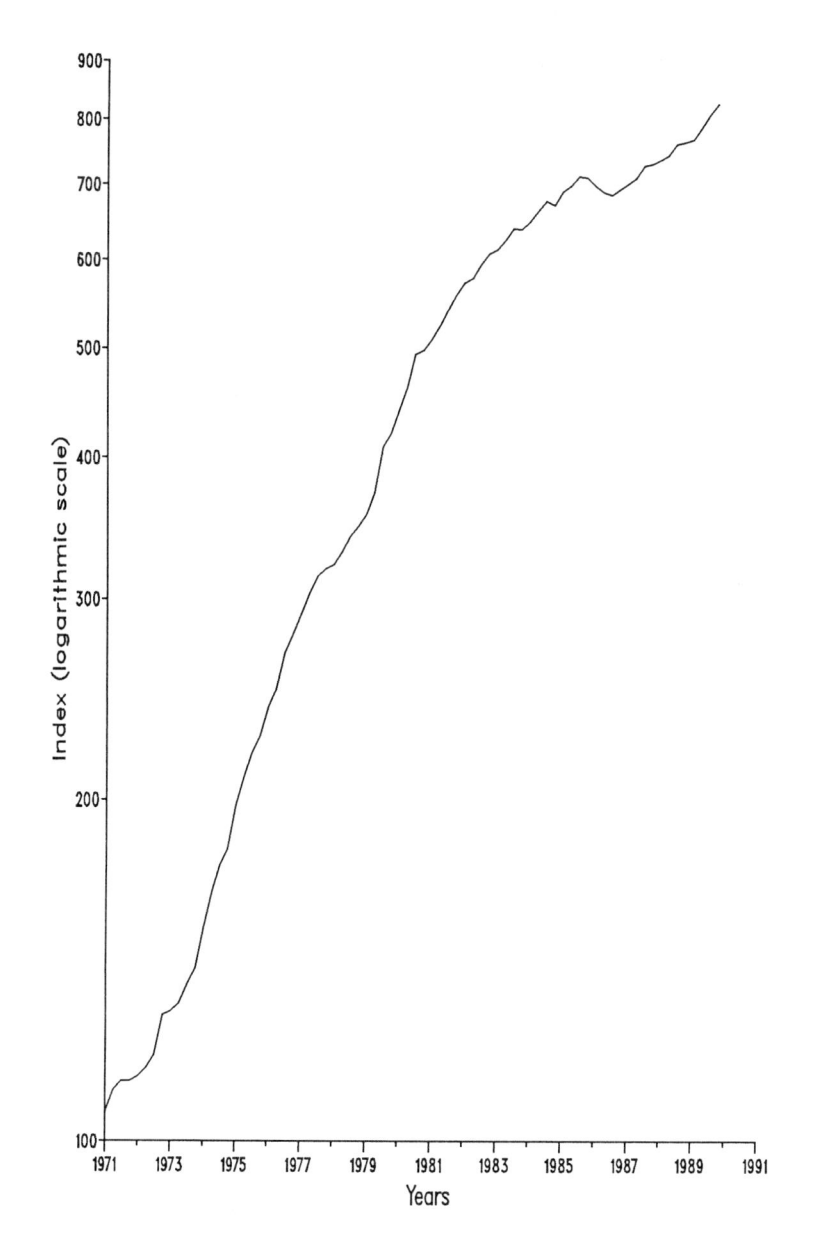

Figure 6.2.4 Spon's Constructed Civil Engineering Cost Index (1970 = 100)
1971 Q1 – 1989 Q4

Table 6.2.5 Spon's Constructed Landscaping (Hard Surfacing and Planting) Cost Index

Base: 1976 = 100

Year	Q1	Q2	Q3	Q4	Average
1976	94	97	102	108	100
1977	111	114	120	121	117
1978	122	125	132	134	128
1979	135	138	155	160	147
1980	163	168	187	188	177
1981	189	192	199	203	196
1982	205	206	215	217	211
1983	217	221	229	229	224
1984	231	233	241	243	237
1985	245	247	256	258	252
1986	262	263	270	273	267
1987	275	278	287	289	282
1988	291	294	303	307	299
1989	308	313	325**	327**	318**
1990	329**				
1991					
1992					
1993					
1994					
1995					

Source: *Spon's Landscape and External Works Price Book.*

** Provisional.

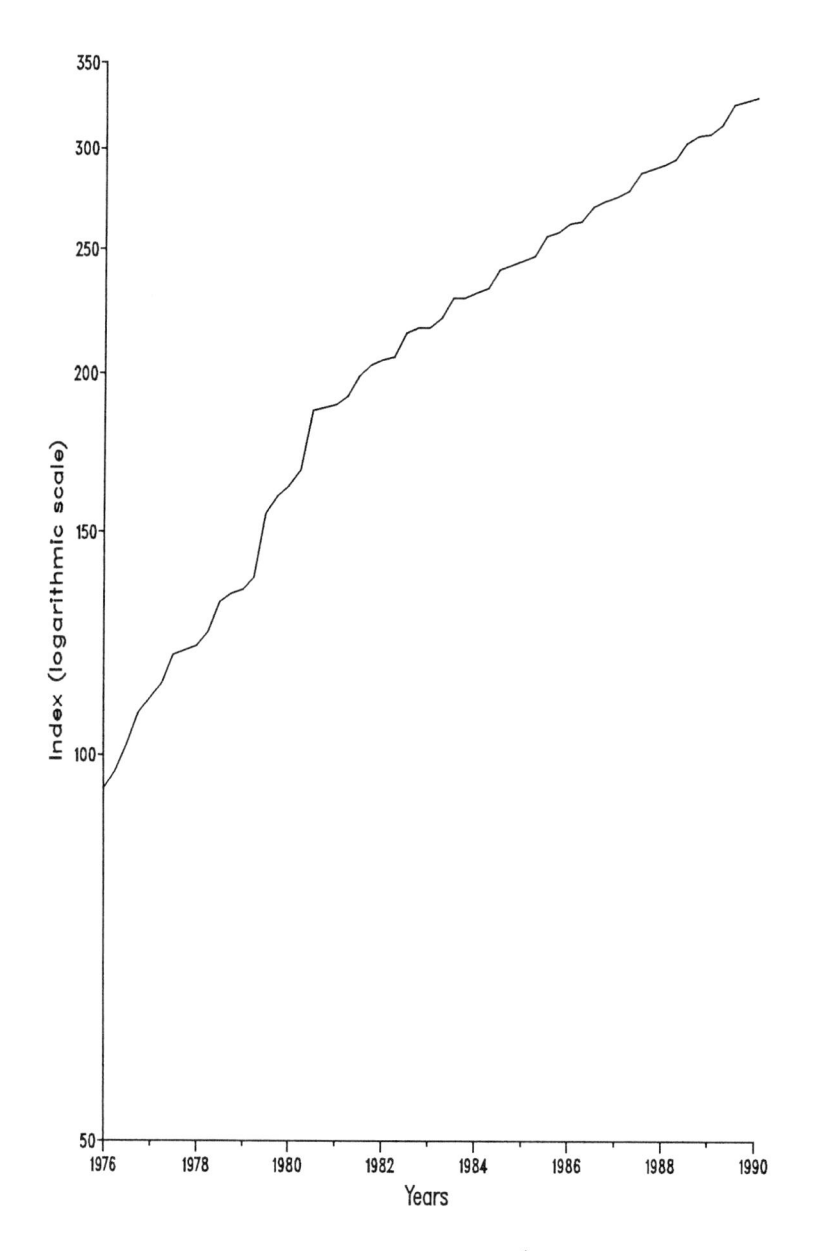

Figure 6.2.5 Spon's Constructed Landscaping (Hard Surface and Planting) Cost Index (1976 = 100), 1976 Q1 – 1990 Q1

6.3 'BUILDING' HOUSING COST INDEX

Type of Index

Building cost index (BCI).

Series: Coverage and Breakdowns

Analyses of new housebuilding in the public and private sectors as follows:

Series	Table Reference
'Building' Housing Cost Index, 1974 to date	6.3.1

Base Dates and Period Covered

December 1973   from 1974 to date.

Frequency

Monthly although the authors have converted the figures in the table that follows to a quarterly series.

Geographical Coverage

United Kingdom.

Type and Source of Data

The inputs to the 'Building' housing cost index are plant and material costs obtained from suppliers and merchants and all-in calculated labour costs based on the CIOB Code of Estimating Practice.

Method of Compilation

The original method of compilation of the 'Building' housing cost index was fully described in the issue of *Building* dated 3 January 1975 and the rebased details were fully described in the issue of *Building* dated 7 April 1978. There have been a few minor amendments since then. It is strictly a measure of the movement of costs to contractors, a factor cost index, and does not include assumptions for market conditions.

House prices are affected by forces other than construction costs, such as demand, availability of mortgages and land prices. Similarly, tender prices are influenced by market conditions. Such imponderables cannot properly be incorporated in a factor cost index, and therefore the 'Building' housing cost index makes no allowance for land prices or the profit element in building or development. An allowance for site and head office overheads has been included based on representative and verified data.

The weightings indicate the relative importance to overall costs of individual items of labour and material. They can be used to estimate the effect on house building costs of known increases in material and labour costs. For example, at the base date of the index (December 1973) bricks represented 3.4 per cent of the total cost of the house. A 20 per cent rise in brick prices would therefore have meant an increase in the cost of the house of 0.68 per cent due to this factor alone (i.e. 3.4 per cent x 20/100 = 0.68 per cent). The quantity weightings for the index were derived for the house and site works by preparing an approximate bill of quantities in the normal way. The bill items were split down into their labour and material contents. These were reabstracted to arrive at a bill of labour and material items. Materials of a similar character were grouped together when possible and quantity weightings derived.

The weightings were priced out and, again where possible, represented by single items for each group of materials. For example, 8 windows of varying sizes were represented by 12 windows of a single size with the same overall cost.

Site overheads were estimated on standard preliminary items priced at current levels. Representative items have been used to monitor changes in the levels of overheads. Initially included in the overheads was an allowance for increased costs of the notional 15 months contract period; this has since been deleted. Off-site overheads have been included at 10 per cent. The house used in the index is a two-storey, three-bedroomed, semi-detached house of traditional construction. As such it is held to be representative of the majority of houses being constructed.

For the purpose of pricing siteworks, 32 units were placed on a notional site of 1.054 ha. It has been assumed that it is a clear site in the suburbs of a large town, has good access, is well positioned for the supply of labour and materials, has plenty of room for storage of materials, and has good bearing soil.

The original specification was as follows, although certain alterations have been made to comply with changes in the Building Regulations.

Specification
Gross floor area
87 square metres
Substructure
    Concrete strip foundations, brick cavity walls with concrete cavity fill. One brick cross wall. Asbestos-based damp proof course, 100mm hardcore bed. 125mm mesh reinforced concrete slab. Polythene damp proof membrane.

Cost indices

Superstructure
 Upper floors: timber joists on joist hangers. Tongued and grooved boarding. Roof: timber trusses at 500mm centre. Concrete interlocking roof tiles. Underfelt. Insulating mat. PVC rainwater goods.
 External walls: 270mm cavity walls. Tudor facings outer skin. Concrete block inner skin. Weather boarding to elevations at first floor level.
 Windows and external doors: standard wood casement windows with OQ sheet glass. Two panelled glazed doors with Georgian wired cast glass.
 Internal walls and partitions: one brick party wall. Timber stud partitions with plasterboard skins.
 Internal doors: hollow-cored hardboard-faced flush doors.
Internal finishes
 Wall finishes: 13 mm two coat carlite plaster to brick and block walls. 5mm skim coat on plasterboard. Emulsion paint.
 Floor finishes: 50mm cement and sand screed.
 Ceiling finishes: 12mm plasterboard, foil backed at first floor. 5mm skim coat. Vinyl paint.
Fittings and furnishings
 Kitchen units. Worktops. Shelving.
Services
 Sanitary appliances: white glazed vitreous china lavatory basins and low level wc suits. Acrylic bath set. Stainless steel sink and drainer. Taps and plastic traps.
 Disposal installation: PVC rainwater down pipes. UPVC waste and soil and vent pipes. Glazed ware drains.
 Water installation: mains supply cold water and hot water service in copper tubing.
 Heat source and space heating: gas fired automatic boiler. Hot water radiators. Hot water cylinder. Pipework in copper.
 Electrical installation: concealed circuit to light fitting and socket outlets.
 Gas installation: supply to boiler only.
External works
 Site works: 150mm mesh reinforced concrete road on hardcore bed. 50mm precast concrete paving and kerbs. Concrete crossovers. Grass seeding to front gardens. Softwood creosoted palisade fencing to site boundary. Strained wire fencing with precast concrete posts between gardens. Framed ledged and braced gates.
 Drainage: glazed vitrified clay (GVC) gullies and pipes. Engineering brick manholes cast iron covers and step irons.
 Minor building works: attached garages concrete strip foundations, reinforced slab on hardcore on polythene dpm. Half-brick walls in facings with attached piers. Two layers bitumen felt on tongued and grooved boarding on softwood joists. PVC gutters and down pipes. Metal garage doors.

The alternations made to comply with the changes in the Building Regulations are the substitution of aerated concrete for the clinker inner skins of cavity walls and the change in thickness from 25mm to 50mm for the roof insulation.
 The base value weightings at February 1978 (total 1000) were as follows:

Building cost house and site works

Building craftsmen	206
Building labourers	189
Plumber and mate	54
Electrician	14
Sand and aggregate	22
Cement	22
Reinforcement	4
Membranes	2
Precast concrete	7
Bricks	46
Blocks	16
Roof tiles	11
Bitumen felt	5
Timber	84
Thermal insulation	3
Doors	10
Windows	12
Kitchen fittings	15
Ironmongery	3
Metalwork	11
Rainwater and waste	4
Sanitary fittings	14
Copper tubes and fittings	12
Boiler	9
Cold water tank	1
Hot water cylinder	3
Radiators	8
Electrical goods	12
Plaster and plasterboard	14
Glazing	6
Paint	6
Drainage goods	9

Site overheads

Supervision	37
Site accommodation	6
Temporary roads	1
Temporary hoardings	1
Temporary telephones	1
Scaffolding	13
Light and power	2
Leaving building clean	1
Transport	1
Plant tools and vehicles	6
Water for the works	6
Insurance	1
Head office overheads	91
	1000

Commentary

The index measures the changes in costs of labour, materials and plant (i.e. basic cost to the contractor) between one time period and another for a typical housing scheme. They *do not* make any allowance for market conditions. Their main use, therefore, is in making an assessment of the change in cost to a contractor, for all or one particular factor, between one time period or another.

Publications

(a) Data Source

Building.

(b) Description of Methodology

Details of the methodology were initially given in *Building*, 3 January 1975, pp.41-42 and the rebased details were given in *Building*, 7 April 1978, p.63.

Table 6.3.1 'Building' Housing Cost Index

Base: Dec 1973 = 100

Year	Q1	Q2	Q3	Q4	Average
1974	101	105	113	117	109
1975	124	130	140	142	134
1976	145	151	160	163	155
1977	168	173	178	181	175
1978	182	185	194	199	190
1979	204	208	231	238	220
1980	244	252	278	281	264
1981	285	290	300	305	295
1982	310	316	327	329	321
1983	330	337	347	349	341
1984	351	355	365	364	359
1985	368	374	385	386	378
1986	387	391	401	403	396
1987	409	414	425	427	419
1988	431	437	452	456	444
1989	462	468	490	495	479
1990	500**				
1991					
1992					
1993					
1994					
1995					

Source: 'Building'.

** Provisional.

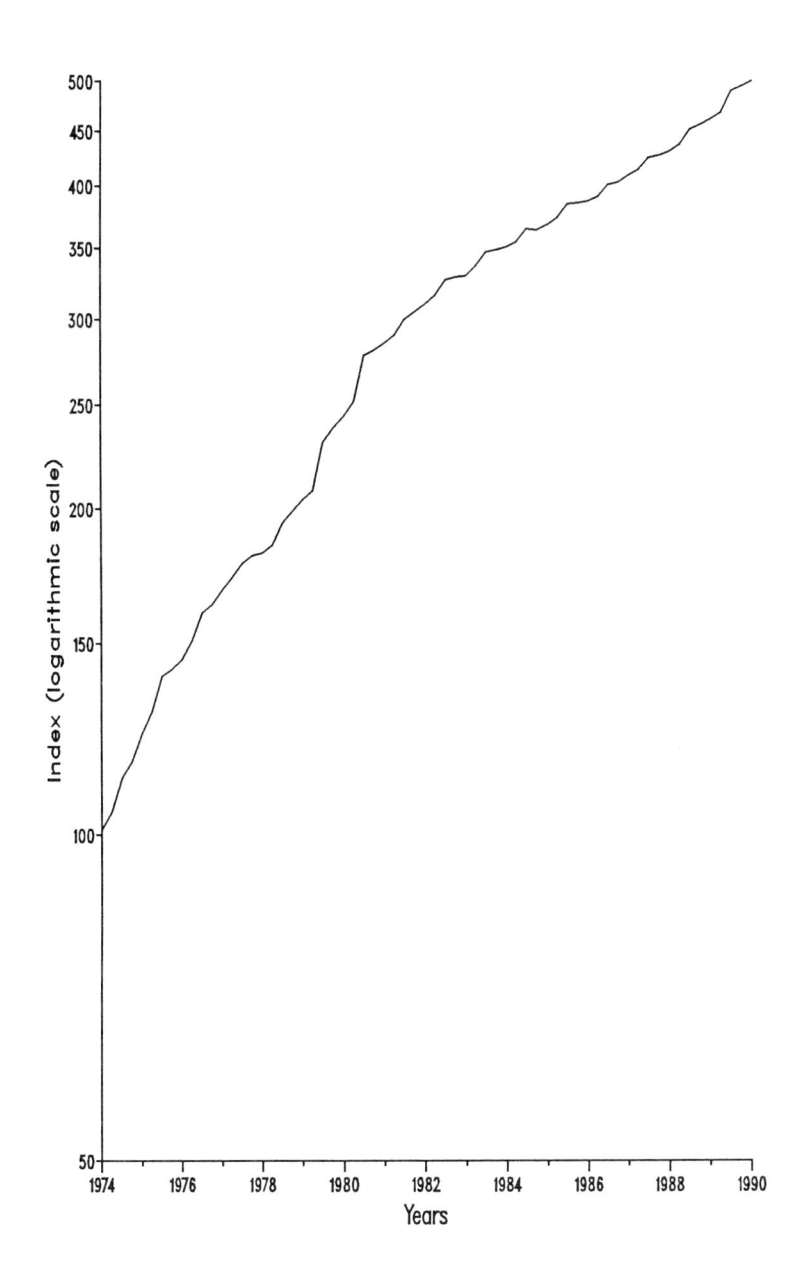

Figure 6.3.1 'Building' Housing Cost Index (Dec 1973 = 100), 1974 Q1 – 1990 Q1

6.4 SCOTTISH OFFICE BUILDING COST INDEX

Type of Index

Building Cost Index (BCI).

Series: Coverage and Breakdowns

Analyses of new building as follows:

Series	*Table Reference*
Scottish Building Cost Index, 1974 to date	6.4.1

Base Dates and Period Covered

1970	from 1970 to 1979
1975	from 1974 to 1983
1980	from 1980 to 1987
1985	from 1980 to date

The pre-1985 series are shown with 1985 = 100 in the tables below (converted arithmetically by the authors).

Frequency

Quarterly.

Geographical Coverage

Scotland.

Type and Source of Data

The inputs to the Scottish Office Building Cost Index are the Work Category Indices (Series 2) prepared by the Property Services Agency for use with their Price Adjustment Formulae for Construction Contracts.

Method of Compilation

The index is compiled by the Scottish Office Building Directorate. It measures the movement of building costs by the application of Series 2 Price

Adjustment Formulae Indices (Series 1 prior to 1979 Q3) to an amalgam of three functional types of building in the Public Sector (traditional housing, district general hospital and primary school) and shows the change in the cost to contractors of labour, materials and plant reflected in the Price Adjustment Formula indices weighted according to the tender amounts of similar work in each contract.

Commentary

The index measures the changes in costs of labour, materials and plant (i.e. basic cost to the contractor) between one time period and another. It *does not* make any allowance for market conditions. Its main use, therefore, is in making an assessment of the change in cost to a contractor between one time period or another.

Publications

(a) Data Source

Statistical Bulletins, quarterly, Housing Statistics Unit, Scottish Development Department, available from Scottish Office Library, Publications Sales, Room 2/65, New St Andrews House, Edinburgh EH1 3TG.

(b) Description of Methodology

The construction, scope and uses of the index are as described above.

Table 6.4.1 Scottish Office Building Cost Index

Base: 1985 = 100*

Year	Q1	Q2	Q3	Q4	Average
1970	16	16	16	16	16
1971	17	17	18	18	18
1972	18	18	18	21	19
1973	21	22	23	24	23
1974	25	27	28	29	27
1975	31	32	34	35	33
1976	36	38	41	42	39
1977	43	45	47	47	46
1978	48	48	51	52	50
1979	53	55	60	62	58
1980	64	66	72	73	69
1981	74	75	77	79	76
1982	81	82	85	86	84
1983	86	88	91	92	89
1984	92	94	97	97	95
1985	98	99	102	101	100
1986	102	102	105	105	104
1987	106	108	111	111	109
1988	112	113	117	118	115
1989	119	121	126		
1990					
1991					
1992					
1993					
1994					
1995					

Source: Scottish Office.

*   Indices from 1970 to 1984 inclusive converted arithmetically by the authors from bases of 1970 = 100, 1975 = 100 and 1980 = 100 to 1985 = 100.

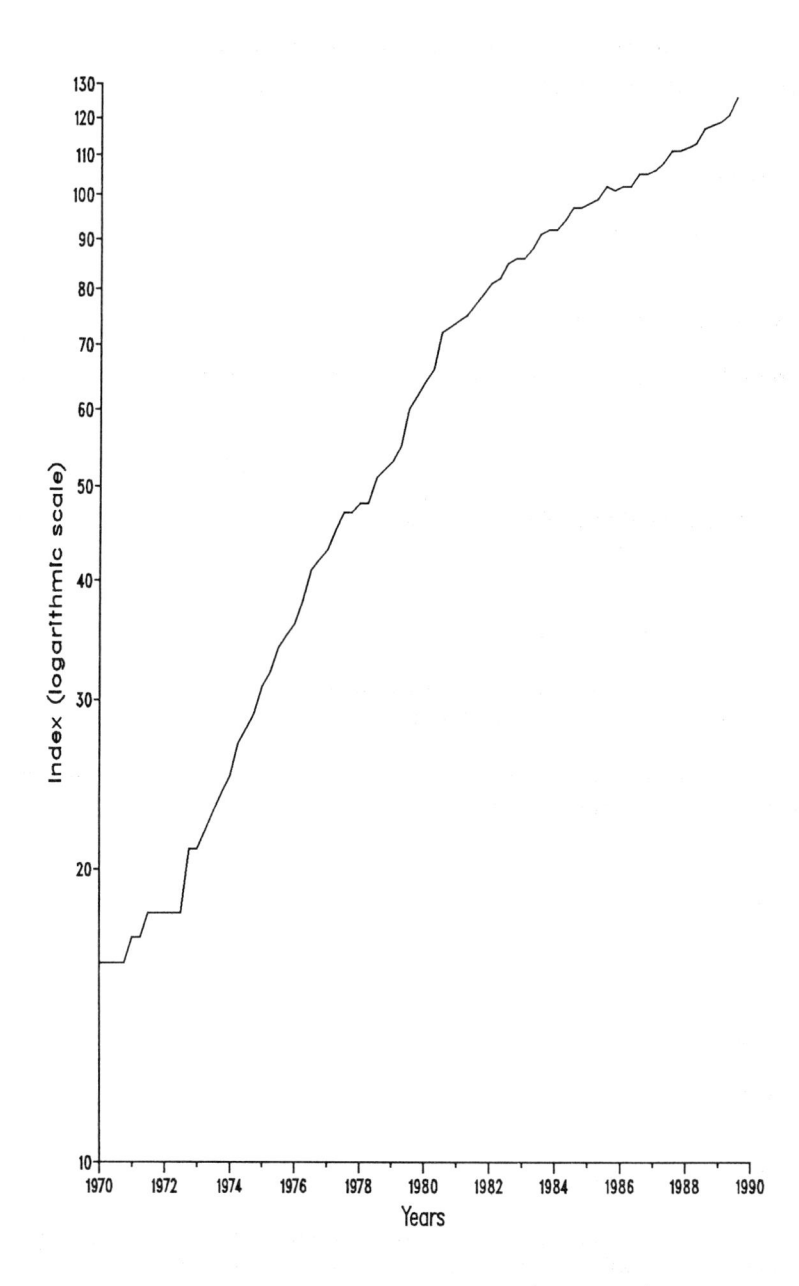

Figure 6.4.1 Scottish Office Housing Cost Index (1985 = 100), 1970 Q1 – 1989 Q3

182 *Spon's handbook of construction cost and price indices*

6.5 ABI/BCIS HOUSE REBUILDING COST INDEX

Type of Index

Building cost index (BCI).

Series: Coverage and Breakdowns

House rebuilding in the public and private sectors as follows:

Series	*Table Reference*
ABI/BCIS House Rebuilding Cost Index, July 1978 to date	6.5.1

Base Dates and Period Covered

| July 1978 | from July 1978 to September 1988 |
| September 1988 | from September 1988 to date. |

The pre-September 1988 series are shown with September 1988 = 100 in the table below (converted arithmetically by the authors).

Frequency

Monthly.

Geographical Coverage

United Kingdom.

Type and Source of Data

The inputs to the ABI/BCIS House Rebuilding cost index are the changes in the costs of labour, material, plant, profit, overheads and fees.

Method of Compilation

The following is an extract from the *Guide to House Rebuilding Costs for Insurance Valuation* prepared on behalf of the Association of British Insurers by the Building Cost Information Service of The Royal Institution of Chartered Surveyors:

'The ABI/BCIS House Rebuilding Cost Index is prepared by the Building Cost Information Service of the Royal Institution of Chartered Surveyors for the Association of British Insurers. The index is intended to update the house rebuilding cost figures published in this Guide, in the ABI's leaflet 'Buildings Insurance for the Home Owner' and for various insurance companies in their renewal notices. The index is based on changes in the rebuilding costs for all building types in Tables 1-4 of this Guide, resulting from changes in the input costs of labour, material, plant, profit, overheads and fees. Prior to September 1982 the index monitored the rebuilding cost of a modern, small, basic quality, semi-detached house. The rebuilding costs are recalculated each month (based on costs current on the third Wednesday of the month).

The index was first calculated in July 1978. From September 1978 until December 1980 it was calculated quarterly at the end of September, December, March and June. From January 1981 it is available on a monthly basis.'

Commentary

The indices measures the changes in costs of labour, materials and plant (i.e. basic cost to the contractor) between one time period and another. However, they *do* try to make an allowance for market conditions in recalculating overheads and profit each time. The use of the index, therefore, is in making an assessment of the change in cost of house rebuilding. Although this index is a factor cost index the attempt to allow for market conditions tends to put it in a category of its own.

Publications

(a) Data Source

The BCIS provides an updating service for the ABI/BCIS house rebuilding cost index which is available on subscription from Building Cost Information Service, 86/87 Clarence Street, Kingston-upon-Thames, Surrey, KT1 1RB. Subscribers to this service receive the monthly index as soon as it is available which is normally within two weeks of the index date. The latest figure is also subsequently published in the *Chartered Surveyor Weekly* and in various insurance publications.

(b) Description of Methodology

Details of the methodology are available from the Building Cost Information Service as above.

Table 6.5.1 ABI/BCIS House Rebuilding Cost Index

Base: September 1988 = 100*

Year	Q1	Q2	Q3	Q4	Average
1978			44.5	45.7	–
1979	47.3**	51.4**	54.5**	55.0**	–
1980	56.9**	63.0**	62.6**	62.9**	–
1981	63.7	64.5	65.2	65.6	64.8
1982	66.4	67.6	70.1	70.6	68.7
1983	71.3	72.3	74.0	74.4	73.0
1984	74.9	76.1	77.9	78.3	76.8
1985	78.8	80.2	82.5	82.6	81.0
1986	82.8	83.7	85.8	86.3	84.7
1987	87.8	89.1	91.3	91.7	90.0
1988	92.3	93.9	98.6	101.2	96.5
1989	102.9	105.0	109.9	111.3	107.3
1990	111.7				
1991					
1992					
1993					
1994					
1995					

Source: BCIS.

*   Indices from July 1978 to August 1988 inclusive converted arithmetically by the authors from base of July 1978 = 100 to September 1988 = 100 using the BCIS conversion factor of 2.296.

**   The figure given from 1978 to 1980 inclusive is the third month figure for each quarter i.e. March, June, September and December.

Note: The figures from 1981 to 1990 are converted from the published monthly figures.

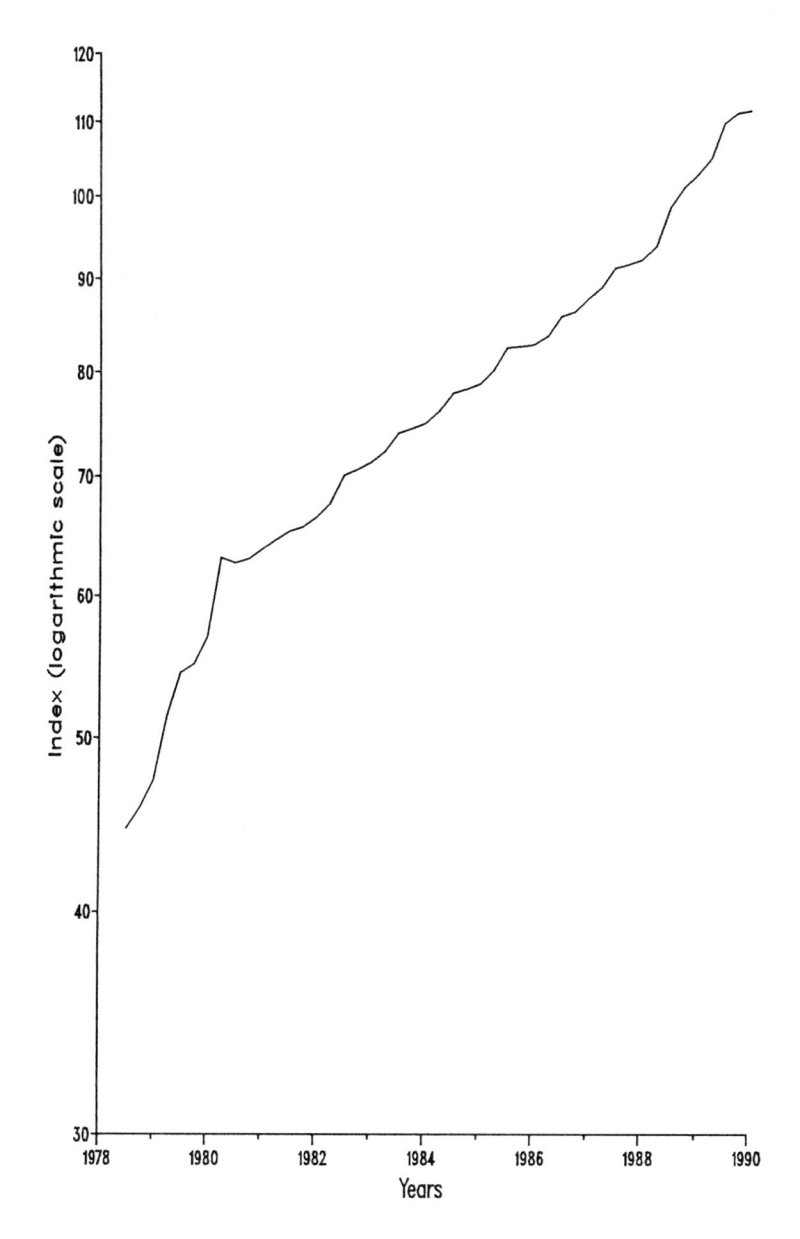

*Figure 6.5.1 ABI/BCIS House Rebuilding Cost Index (Sept 1988 = 100)
1978 Q3 – 1990 Q1*

6.6 ASSOCIATION OF COST ENGINEERS - INDICES OF ERECTED PLANT COSTS

Type of Index

Factor Cost Index (FCI).

Series: Coverage and Breakdowns

Analyses of new chemical, petro-chemical and petroleum projects with breakdowns by type as follows:

Series	*Table Reference*
Association of Cost Engineers Indices of Erected Plant Costs	
Plant A, 1975 to date	6.6.1
Plant B, 1975 to date	6.6.2
Plant C, 1975 to date	6.6.3
Plant D, 1975 to date	6.6.4

Base Dates and Period Covered

1958	from 1958 to 1977
1970	from 1977 to 1979
1975	from 1975 to 1988
1985	from 1980 to date

The pre-1985 series are shown with 1985 = 100 in the tables below (converted arithmetically by the authors).

Frequency

Monthly.

Geographical Coverage

United Kingdom.

Type and Source of Data

The inputs to the Association of Cost Engineers Indices of Erected Plant Costs are based on four typical plants and are factor costs as shown below.

Method of Compilation

The basis of calculation of the Association of Cost Engineers Indices of Erected Plant Costs was explained in an article in *The Cost Engineer* in March 1964 and updated index figures have been included in subsequent editions of that journal, which are obtainable from the Secretary of the Association, 33 Ovingdon Square, London SW3 1LJ.

The indices are related to chemical, petrochemical and petroleum projects in the United Kingdom and reflect the major changes in cost to contractors but not changes in tender levels. In addition to changes in labour and material costs, tenders are affected by other factors such as the degree of competition in the particular industry and area where the work is to be carried out, the availability of labour and the general economic situation. This has meant in recent years that, when there has been an abundance of work, tender levels have often increased at a greater rate than can be accounted for by increases in basic labour and material costs and, conversely, when there is a shortage of work this has often resulted in keener tenders. Allowances for these factors are impossible to assess on a general basis and can only be based on experience and knowledge of the particular circumstances.

A revision of the Department of Trade and Industry's productivity index is taken into account from the first quarter of 1969 onwards.

The indices are based on four typical plants lettered A, B, C and D. The essential contents of the four plants can be gauged from the following two tables, which show the cost proportions of the major elements as they were during the original base year (1958).

Weightings of major elements converted to 1975

Category of cost	Plant A %	Plant B %	Plant C %	Plant D %
Mechanical and electrical material and equipment	50.3	54.4	52.6	57.2
Erection labour	31.4	29.3	30.4	26.9
Civil and building materials	4.8	5.0	4.4	5.6
Administrative, technical and clerical salaries	10.2	8.5	9.5	7.8
Construction equipment/transport	3.3	2.8	3.1	5.2
	100.0	100.0	100.0	100.0

Note: Contractor's overheads and profit, clients costs and site preparation are excluded.

Cost indices 189

Proportions of sub-elements within mechanical and electrical material and equipment in 1975

Component	Plant A %	Plant B %	Plant C %	Plant D %
Constructional steelwork	6.1	4.0	4.8	8.3
Heat exchange equipment	11.9	17.5	25.9	-
Above ground piping fittings and valves	21.3	15.8	21.0	9.2
Miscellaneous machinery	5.4	8.8	-	27.2
Pumps and drivers	7.8	4.7	4.6	9.8
Compressors and drivers	3.9	-	1.8	-
Tanks and vessels	16.2	26.4	20.5	32.8
Fired heaters	12.9	8.4	-	-
Underground piping fittings and valves	1.7	0.8	0.8	0.9
Process instrumentation	5.1	8.0	11.5	4.1
Electrical machinery	6.8	3.8	7.3	6.8
Thermal insulation	0.9	1.8	1.8	0.9
	100.0	100.0	100.0	100.0

Commentary

The indices measure the changes in costs of labour, materials and plant (i.e. basic cost to the contractor) between one time period and another. They *do not* make any allowance for market conditions. Their main use, therefore, is in making an assessment of the change in cost to a contractor, for large industrial projects, between one time period or another.

Publications

(a) Data Source

The *Cost Engineer* published by the Association of Cost Engineers, 33 Ovingdon Square, London SW3 1LJ.
Spon's Mechanical and Electrical Price Book, 1989.

(b) Description of Methodology

The *Cost Engineer*, March 1964 and *Spon's Mechanical and Electrical Price Book*, 1989.

Table 6.6.1 Association of Cost Engineers - Erected Plant Cost Index - Plant A

Base: 1985 = 100*

Year	End of Quarter				Average
	March	June	September	December	
1975	31.4	33.2	34.7	35.6	
1976	37.4	38.6	40.0	40.9	
1977	42.4	43.7	45.0	45.8	
1978	48.0	49.4	51.1	51.8	
1979	54.5	56.0	59.1	60.3	
1980	64.1	66.1	68.0	69.7	
1981	71.8	73.6	74.7	75.4	
1982	78.7	80.5	81.6	82.8	
1983	83.6	85.3	86.8	88.5	
1984	90.4	92.5	93.7	94.8	
1985	98.0	99.9	101.5	102.8	100.0
1986	104.1	105.7	106.8	107.8	105.7
1987	109.7	111.0	112.2	114.2	111.2
1988	116.5	119.1	121.0	122.5	119.1
1989	126.6	130.8	132.4**		
1990					
1991					
1992					
1993					
1994					
1995					

Source: Association of Cost Engineers.

* Indices from 1975 to March 1985 inclusive converted arithmetically by the authors from base of 1975 = 100 to 1985 = 100.

** Provisional

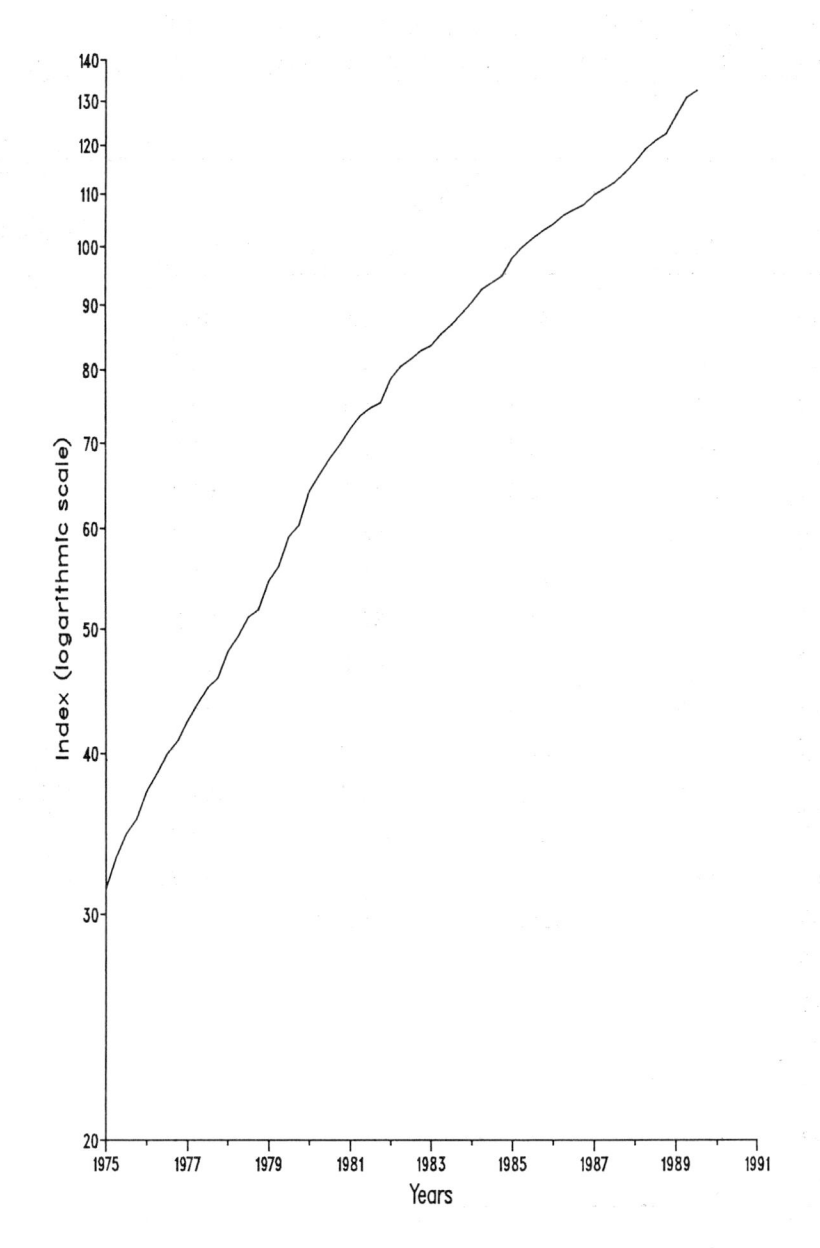

Figure 6.6.1 Association of Cost Engineers – Erected Plant Cost Index – Plant A (1985 = 100), March 1975 – September 1989

Table 6.6.2 Association of Cost Engineers - Erected Plant Cost Index - Plant B

Base: 1985 = 100*

Year	End of Quarter				Average
	March	June	September	December	
1975	32.4	34.2	35.9	36.6	
1976	38.4	39.7	41.1	42.1	
1977	43.5	44.7	46.1	46.9	
1978	49.0	50.4	52.0	52.5	
1979	55.2	56.6	59.7	61.0	
1980	64.7	66.9	68.6	70.3	
1981	72.3	73.9	75.1	76.1	
1982	79.3	81.1	82.2	83.3	
1983	84.1	85.7	87.2	88.8	
1984	90.8	92.6	93.8	94.9	
1985	98.5	99.9	101.4	102.5	100.0
1986	103.7	105.1	106.1	107.3	105.2
1987	109.4	110.7	111.7	113.7	110.8
1988	115.8	118.3	120.3	121.7	118.4
1989	125.3	129.5	130.5**		
1990					
1991					
1992					
1993					
1994					
1995					

Source: Association of Cost Engineers.

*   Indices from 1975 to March 1985 inclusive converted arithmetically by the authors from base of 1975 = 100 to 1985 = 100.

**   Provisional

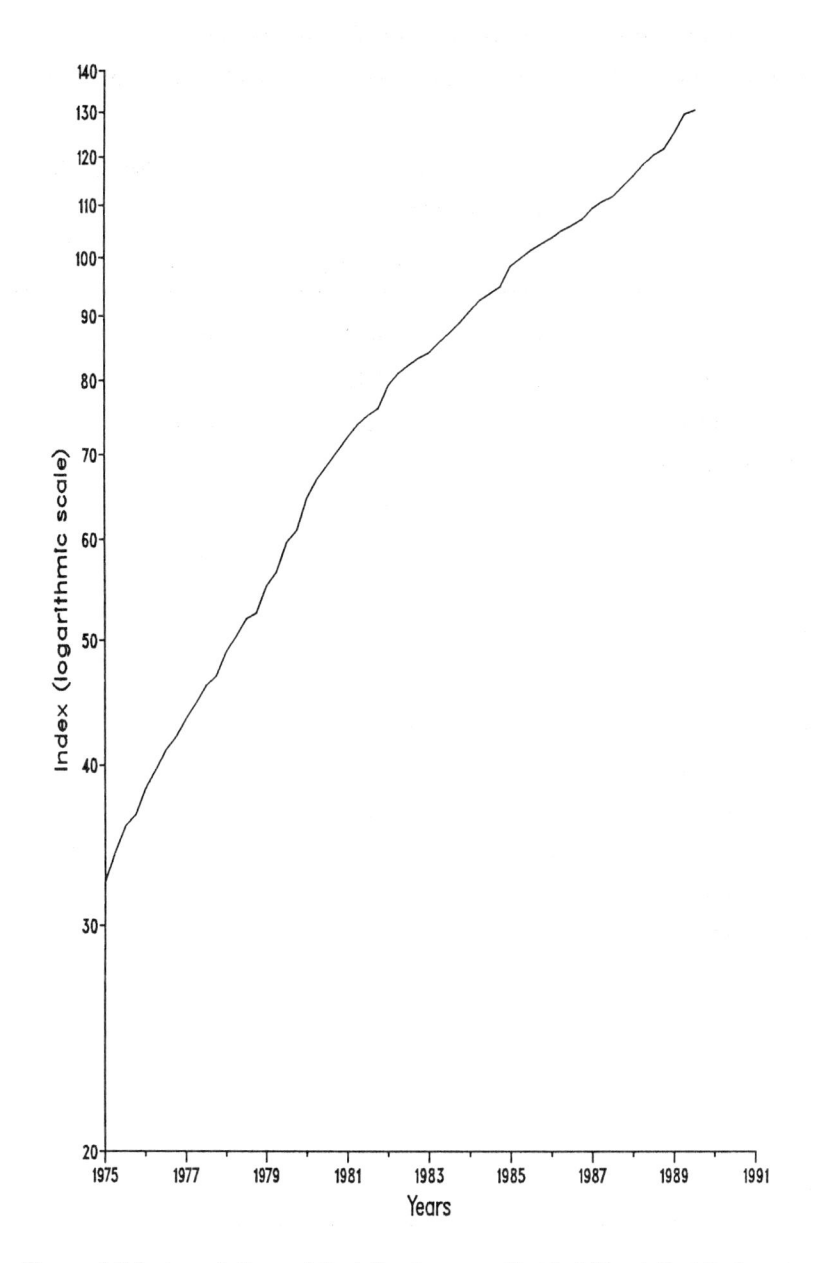

Figure 6.6.2 Association of Cost Engineers – Erected Plant Cost Index – Plant B (1985 = 100), March 1975 – September 1989

Table 6.6.3 Association of Cost Engineers - Erected Plant Cost Index - Plant C

Base: 1985 = 100*

Year	March	June	September	December	Average
1975	32.3	34.4	36.0	36.6	
1976	38.3	39.6	41.0	41.9	
1977	43.2	44.3	45.8	46.6	
1978	48.5	50.0	51.5	51.8	
1979	54.5	55.9	58.8	59.9	
1980	63.5	65.8	67.5	69.1	
1981	70.9	72.5	73.7	74.8	
1982	77.7	79.5	80.9	81.7	
1983	82.5	83.9	85.5	87.1	
1984	89.4	91.1	92.5	93.5	
1985	97.5	99.9	101.4	102.5	100.0
1986	103.4	104.7	105.8	106.9	104.9
1987	109.2	110.5	111.7	113.9	110.7
1988	116.0	118.5	120.5	122.0	118.6
1989	124.8	129.0	130.5**		
1990					
1991					
1992					
1993					
1994					
1995					

Source: Association of Cost Engineers.

*   Indices from 1975 to March 1985 inclusive converted arithmetically by the authors from base of 1975 = 100 to 1985 = 100.

** Provisional

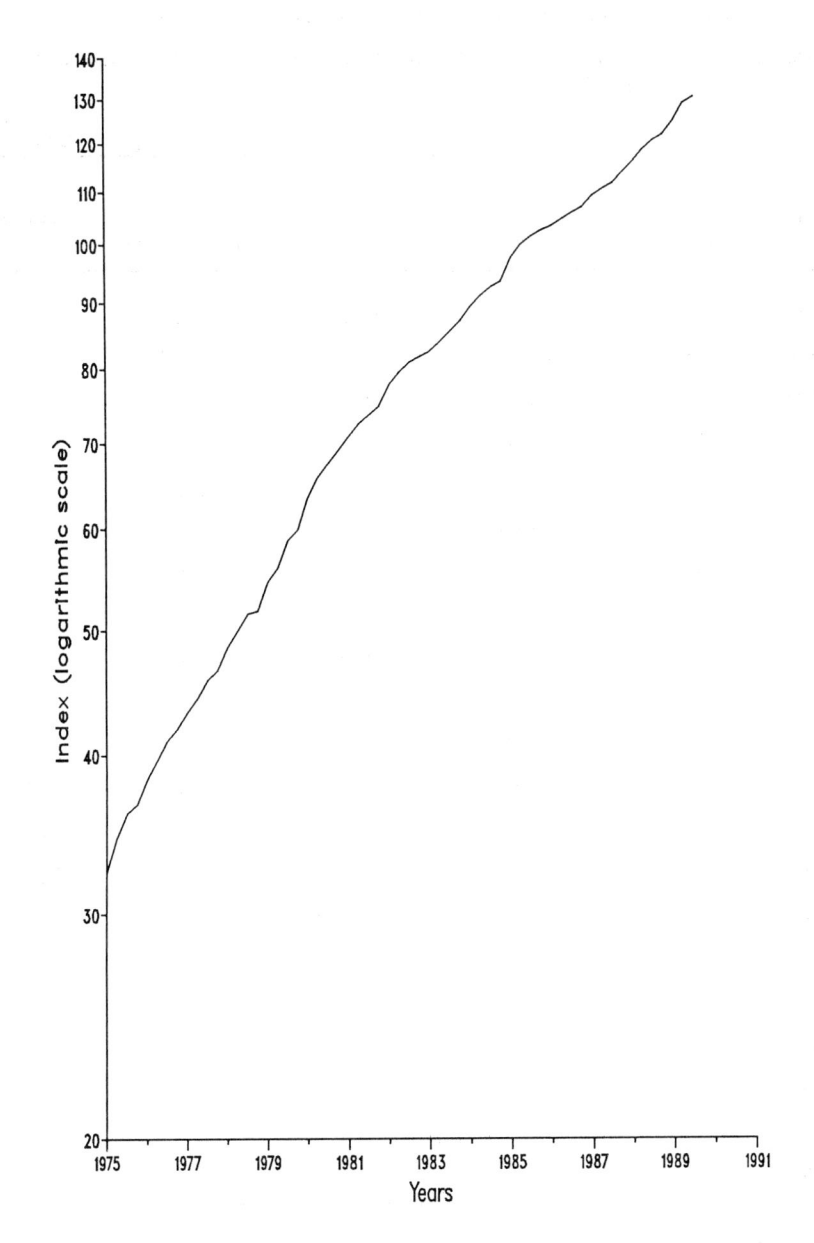

Figure 6.6.3 Association of Cost Engineers – Erected Plant Cost Index –
Plant C (1985 = 100), March 1975 – September 1989

Table 6.6.4 Association of Cost Engineers - Erected Plant Cost Index - Plant D

Base: 1985 = 100*

Year	March	June	September	December	Average
1975	32.9	34.2	35.8	37.1	
1976	38.9	40.1	41.6	42.7	
1977	44.2	45.4	46.5	47.6	
1978	49.7	50.9	52.6	53.9	
1979	55.9	57.6	60.6	62.2	
1980	65.9	68.0	69.7	71.4	
1981	73.4	74.9	76.3	76.9	
1982	80.1	81.9	83.1	84.2	
1983	85.1	86.5	88.0	89.0	
1984	91.0	92.4	93.5	94.6	
1985	97.0	100.0	101.4	102.5	100.0
1986	104.3	105.6	106.7	108.0	105.7
1987	109.9	111.3	112.3	114.2	111.4
1988	116.3	118.5	120.5	122.1	118.7
1989	125.2	128.8	130.2**		
1990					
1991					
1992					
1993					
1994					
1995					

End of Quarter

Source: Association of Cost Engineers.

* Indices from 1975 to March 1985 inclusive converted arithmetically by the authors from base of 1975 = 100 to 1985 = 100.

** Provisional

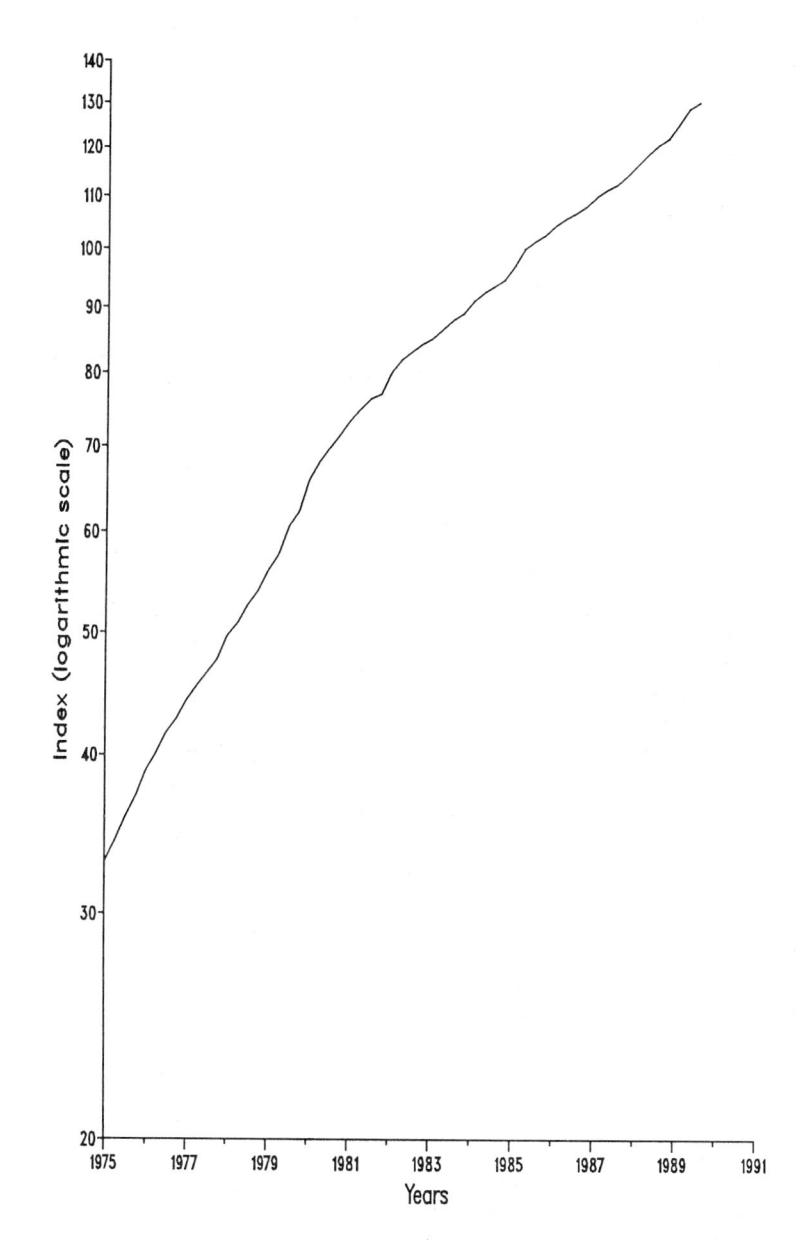

Figure 6.6.4 Association of Cost Engineers – Erected Plant Cost Index – Plant D (1985 = 100), March 1975 – September 1989

6.7 BMI BUILDING MAINTENANCE COST INDICES

Type of Index

Building cost index (BCI).

Series: Coverage and Breakdowns

Analyses of maintenance costs of aspects of buildings in the health service, local authorities and private sector with breakdowns as follows:

Series	Table Reference
BMI Health Service Maintenance Cost Index	
Redecorations, 1980 to date	6.7.1 (BMI 1.1)
Fabric Maintenance, 1980 to date	6.7.2 (BMI 1.2)
Services Maintenance, 1980 to date	6.7.3 (BMI 1.3)
General Maintenance, 1980 to date	6.7.4 (BMI 1.4)
BMI Local Authority Maintenance Cost Index	
Redecorations, 1980 to date	6.7.5 (BMI 2.1)
Fabric Maintenance, 1980 to date	6.7.6 (BMI 2.2)
Services Maintenance, 1980 to date	6.7.7 (BMI 2.3)
General Maintenance, 1980 to date	6.7.8 (BMI 2.4)
BMI Private Sector Maintenance Cost Index	
Redecorations, 1980 to date	6.7.9 (BMI 3.1)
Fabric Maintenance, 1980 to date	6.7.10 (BMI 3.2)
Services Maintenance, 1980 to date	6.7.11 (BMI 3.3)
General Maintenance, 1980 to date	6.7.12 (BMI 3.4)
BMI All-In Maintenance Cost Index	
Redecorations, 1980 to date	6.7.13 (BMI 4.1)
Fabric Maintenance, 1980 to date	6.7.14 (BMI 4.2)
Services Maintenance, 1980 to date	6.7.15 (BMI 4.3)
General Maintenance, 1980 to date	6.7.16 (BMI 4.4)
BMI Maintenance Materials Cost Index	
Redecorations, 1980 to date	6.7.17 (BMI 5.1)
Fabric Maintenance, 1980 to date	6.7.18 (BMI 5.2)
Services Maintenance, 1980 to date	6.7.19 (BMI 5.3)
General Maintenance, 1980 to date	6.7.20 (BMI 5.4)
BMI Cleaning Cost Index, 1980 to date	6.7.21 (BMI 7.1)
BMI Cleaning Materials Cost Index, 1980 to date	6.7.22 (BMI 7.2)
BMI Energy Cost Index, 1980 to date	6.7.23 (BMI 8.1)

Base Dates and Period Covered

1980 Q1   from 1980 to 1990
1990 Q1   (BMI are currently rebasing these indices to this base date).

Frequency

Quarterly.

Geographical Coverage

England.

Type and Source of Data

The inputs to the BMI maintenance cost indices are, for most materials, the 'Producer Price Indices' Table 4, prepared by the Central Statistical Office. These are output price indices based on home sales of manufactured products and imported materials. For some items where CSO indices are not available BMI has prepared its own indices from prices of a sample of appropriate materials suppliers. The labour inputs are monitored using changes in the appropriate wage awards, plus-rates and employers' direct costs for various indices. For private sector workers the BCIS Daywork Standard Hourly Base Rates have been used and for public sector operatives the nationally agreed wage awards have been used. The energy indices for the individual fuels are based on average prices for fuel monitored by the Department of Energy. The information is based on industrial consumers and relates to the average price paid in each quarter. A full list of the input indices is available from Building Maintenance Information Limited, 86/87 Clarence Street, Kingston-upon-Thames, Surrey, KT1 1RB.

Method of Compilation

The BMI maintenance cost indices have been prepared primarily for adjusting the relevant sections of the BMI occupancy cost analyses and cover maintenance, cleaning and energy costs. The indices are base-weighted Laspeyre's indices. Patterns of expenditure in the base period are used and updated for price changes only for each of the input items to produce a measure of overall price change in the index.

The series for maintenance consists of:-

1. Redecorations
2. Fabric Maintenance
3. Services Maintenance
4. General Maintenance

All of the foregoing are related to Health Service work, Local Authority work and the Private Sector, together with an 'All-in' series. The element indices (i.e. Redecoration, Fabric and Services Maintenance) are compiled from weightings of labour and material. The General Maintenance index consists of a weighted average of the three element indices. The weightings for an index which attempts to chart average movements in general maintenance costs can only give guidance on the order of the costs for any specific building in any specific year. It should be noted, for instance, that the level of Services Maintenance contained in the index makes it less applicable to housing than to other classes of buildings. However, the index does give a guide to the movement in maintenance costs and with competent interpretation should assist in comparing expenditure in different years and in updating budgets.

The Cleaning Cost Index has been prepared primarily for adjusting the cleaning section of the BMI occupancy cost analyses. The section covers window cleaning, cleaning of external surfaces and internal cleaning including floors, carpets and dusting and cleaning ledges, furniture and fittings. The weighting system used in the index attempts to model average circumstances so that the index will act as a general guide to movements in cleaning costs. It is therefore unlikely that the make up of cleaning costs for any specific building will exactly fit the model used in the index. The precise breakdown of cleaning expenditure used is available from Building Maintenance Information Limited as above.

The Cleaning Materials Cost Index is the materials input of the Cleaning Cost Index. The Energy Cost Index attempts to chart average movements in energy costs and act as a general guide to these movements. It is unlikely that the make up of energy expenditure on any specific building will exactly fit the model used in the index. For this reason the indices derived from Department of Energy statistics for individual fuels which are used as inputs to the overall Energy Index are also published in the *BMI Quarterly Cost Briefing*. The weighting used in the index is based on a BMI study of Energy Costs, which in turn is based on energy analyses submitted by subscribers to BMI. The precise breakdown used is available from Building Maintenance Information Limited as above.

Commentary

The indices are prepared, primarily, to update the information in the BMI Property Occupancy Cost Analyses but have found many other uses since their introduction. The indices attempt to chart the general movements in occupancy costs over time and should be used accordingly.

Publications

(a) Data Source

BMI Quarterly Briefing of BMI Indices.
Available on subscription from Building Maintenance Information Limited, 86/87 Clarence Street, Kingston-upon-Thames, Surrey, KT1 1RB.

(b) Description of Methodology

Full details of the methodology are available from Building Maintenance Information Limited. The description given above under 'method of compilation' is an edited version thereof reproduced herein by kind permission of Building Maintenance Information Limited.

Table 6.7.1 BMI Health Service Maintenance Cost Index - Redecorations (BMI Table 1.1)

Base: 1980 Q1 = 100

Year	Q1	Q2	Q3	Q4	Average
1980	100.1	100.8	101.3	101.3	100.9
1981	124.4	124.5	124.6	124.7	124.5
1982	133.0	133.3	132.4	131.9	132.7
1983	141.1	141.6	141.5	141.6	141.5
1984	141.6	148.1	148.3	148.2	146.6
1985	148.3	153.9	154.2	152.7	152.3
1986	152.7	161.8	161.8	162.3	159.7
1987	162.6	182.1	182.5	182.6	177.5
1988	182.9	192.2	192.6	193.4	190.3
1989	193.6	205.4	205.6**		
1990					
1991					
1992					
1993					
1994					
1995					

Source: BMI.

** Provisional.

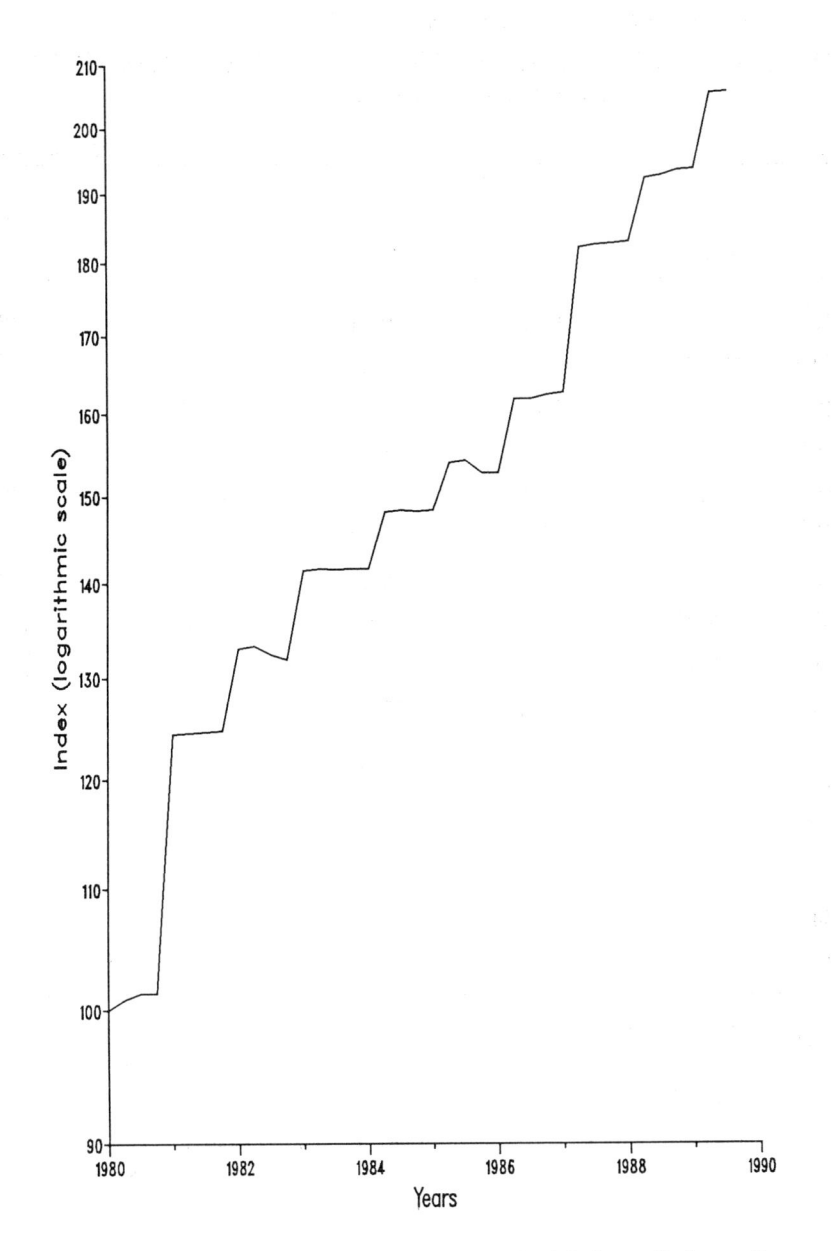

Figure 6.7.1 BMI Health Service Maintenance Cost Index – Redecorations (1980 Q1 = 100), 1980 Q1 – 1989 Q3

Table 6.7.2 BMI Health Service Maintenance Cost Index - Fabric Maintenance (BMI Table 1.2)

Base: 1980 Q1 = 100

Year	Q1	Q2	Q3	Q4	Average
1980	100.0	102.0	103.1	103.6	102.2
1981	120.7	122.1	122.6	123.1	122.1
1982	129.7	130.6	130.4	130.5	130.3
1983	137.9	139.0	139.6	140.3	139.2
1984	141.1	146.7	147.3	148.2	145.8
1985	149.5	154.0	154.3	153.2	152.8
1986	153.0	160.0	160.4	161.4	158.7
1987	162.7	177.5	178.2	178.6	174.3
1988	179.4	187.1	188.4	189.2	186.0
1989	190.3	199.4	200.4**		
1990					
1991					
1992					
1993					
1994					
1995					

Source: BMI.

** Provisional.

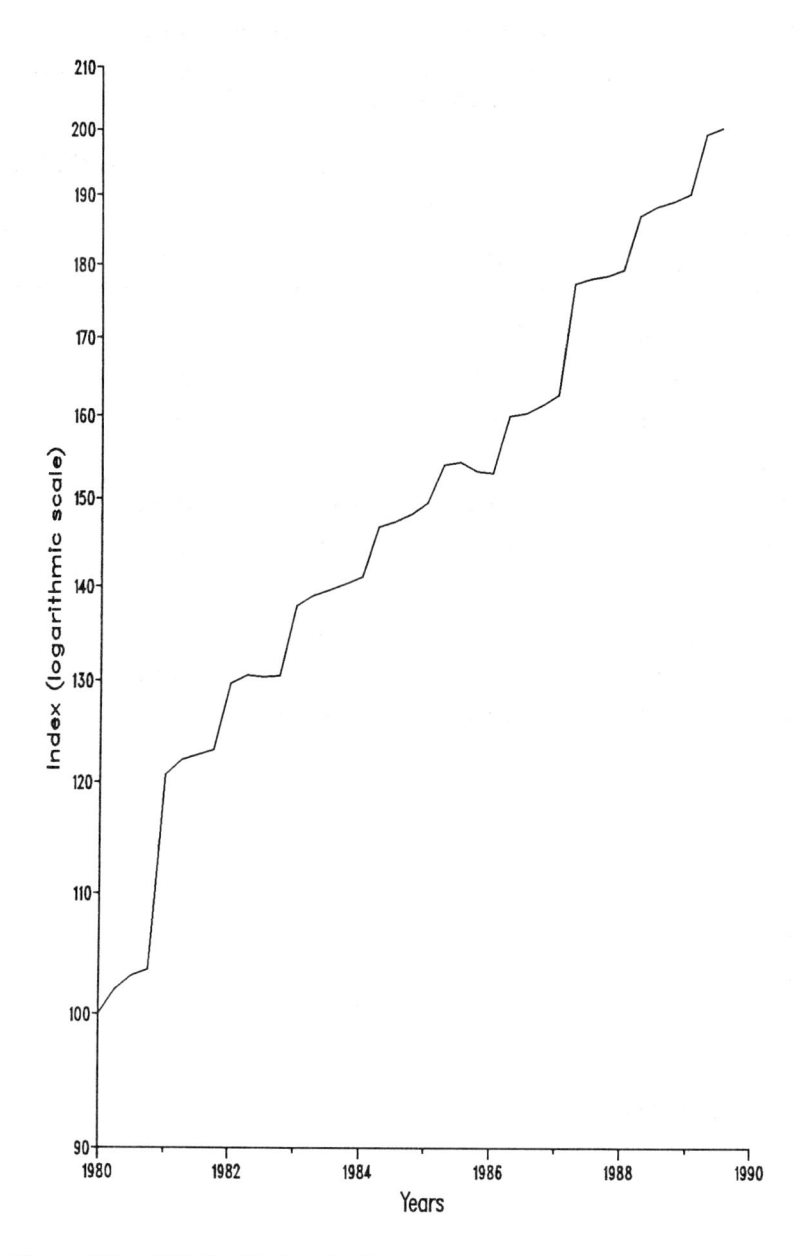

Figure 6.7.2 BMI Health Service Maintenance Cost Index – Fabric Maintenance (1980 Q1 = 100), 1980 Q1 – 1989 Q3

Table 6.7.3 BMI Health Service Maintenance Cost Index - Services Maintenance (BMI Table 1.3)

Base: 1980 Q1 = 100

Year	Q1	Q2	Q3	Q4	Average
1980	100.0	101.1	101.3	101.8	101.0
1981	118.9	119.5	120.1	121.1	119.9
1982	127.8	128.8	128.3	128.5	128.4
1983	137.7	138.8	138.9	138.9	138.6
1984	139.8	146.2	146.4	146.3	144.7
1985	147.7	147.5	151.0	153.1	149.8
1986	153.7	153.6	158.1	161.2	156.7
1987	173.5	179.7	184.2	187.4	181.2
1988	188.8	189.2	194.9	198.6	192.9
1989	199.9	200.4	209.8**		
1990					
1991					
1992					
1993					
1994					
1995					

Source: BMI.

** Provisional.

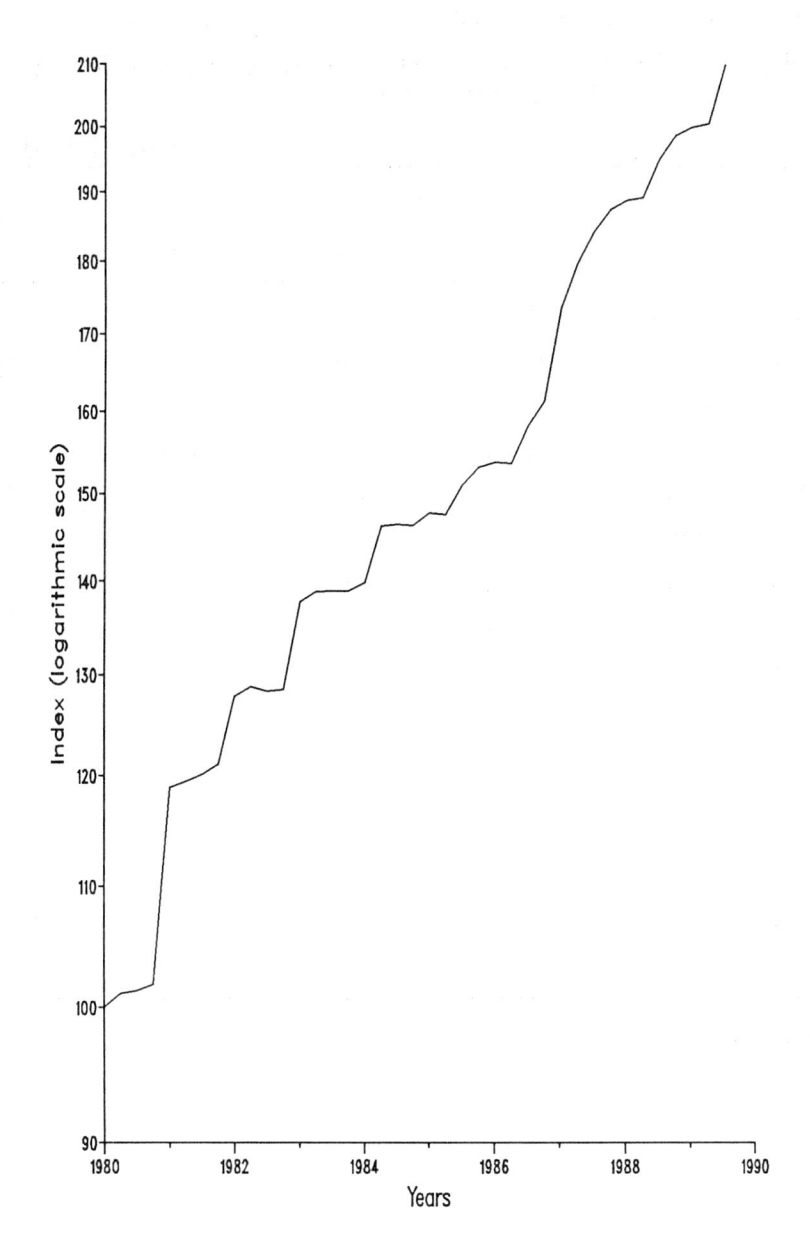

Figure 6.7.3 BMI Health Service Maintenance Cost Index – Services Maintenance (1980 Q1 = 100), 1980 Q1 – 1989 Q3

Table 6.7.4 BMI Health Service Maintenance Cost Index - General Maintenance (BMI Table 1.4)

Base: 1980 Q1 = 100

Year	Q1	Q2	Q3	Q4	Average
1980	100.0	101.3	101.8	102.2	101.3
1981	121.2	121.8	121.3	122.9	121.8
1982	130.0	130.7	130.3	130.2	130.3
1983	139.0	140.0	140.0	140.2	139.8
1984	140.8	147.0	147.3	147.5	145.7
1985	148.4	151.5	153.1	152.9	151.5
1986	153.2	158.2	160.0	161.6	158.3
1987	166.8	179.8	181.9	183.1	177.9
1988	184.1	189.5	192.2	194.1	190.0
1989	195.0	201.7	205.6**		
1990					
1991					
1992					
1993					
1994					
1995					

Source: BMI.

** Provisional.

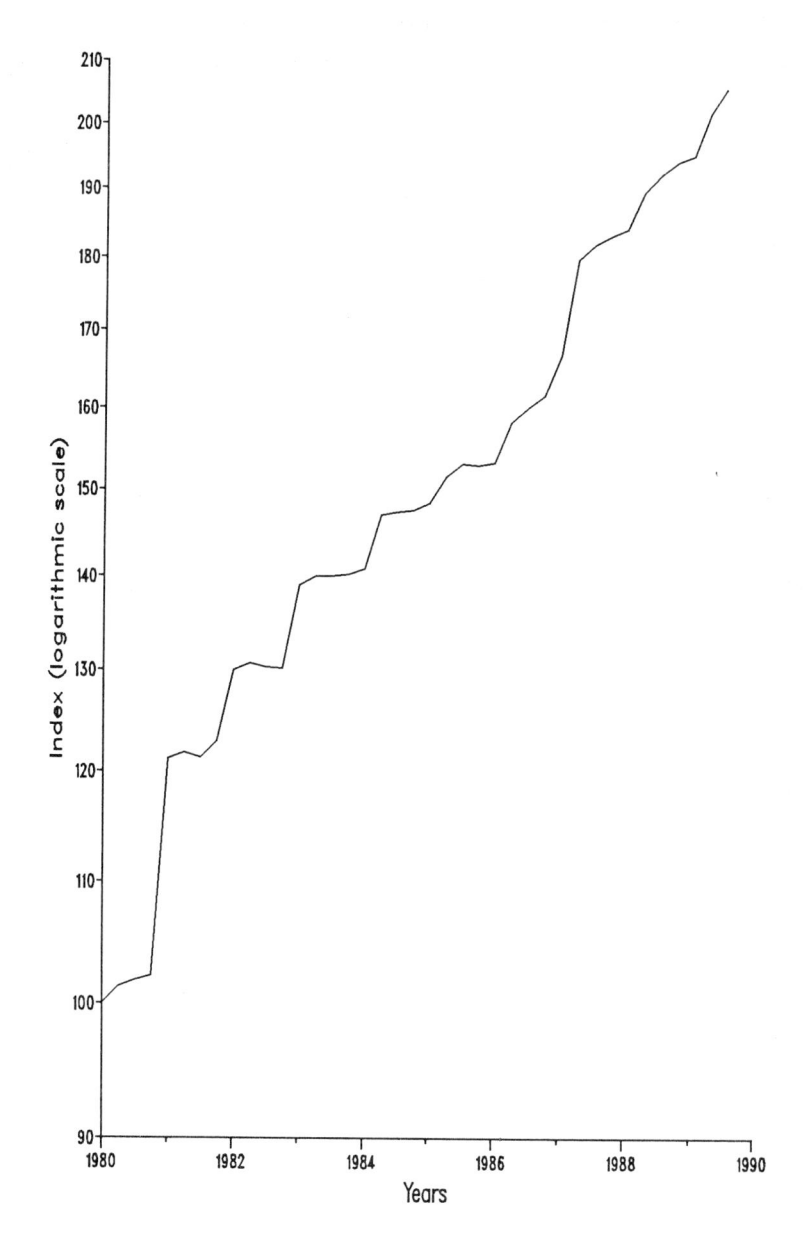

Figure 6.7.4 BMI Health Service Maintenance Cost Index – General Maintenance (1980 Q1 = 100), 1980 Q1 – 1989 Q3

Table 6.7.5 BMI Local Authority Maintenance Cost Index - Redecorations (BMI Table 2.1)

Base: 1980 Q1 = 100

Year	Q1	Q2	Q3	Q4	Average
1980	100.0	100.8	104.4	115.7	105.2
1981	118.4	118.5	118.6	122.7	119.6
1982	124.9	125.2	125.4	130.6	126.5
1983	130.9	131.4	131.4	134.5	132.1
1984	136.2	136.3	136.5	141.4	137.6
1985	143.9	143.1	143.4	141.9	143.1
1986	149.3	151.7	151.2	151.8	151.0
1987	157.1	160.0	164.9	174.0	164.0
1988	174.3	174.4	177.9	184.8	177.9
1989	185.0	185.3	190.4**		
1990					
1991					
1992					
1993					
1994					
1995					

Source: BMI.

** Provisional.

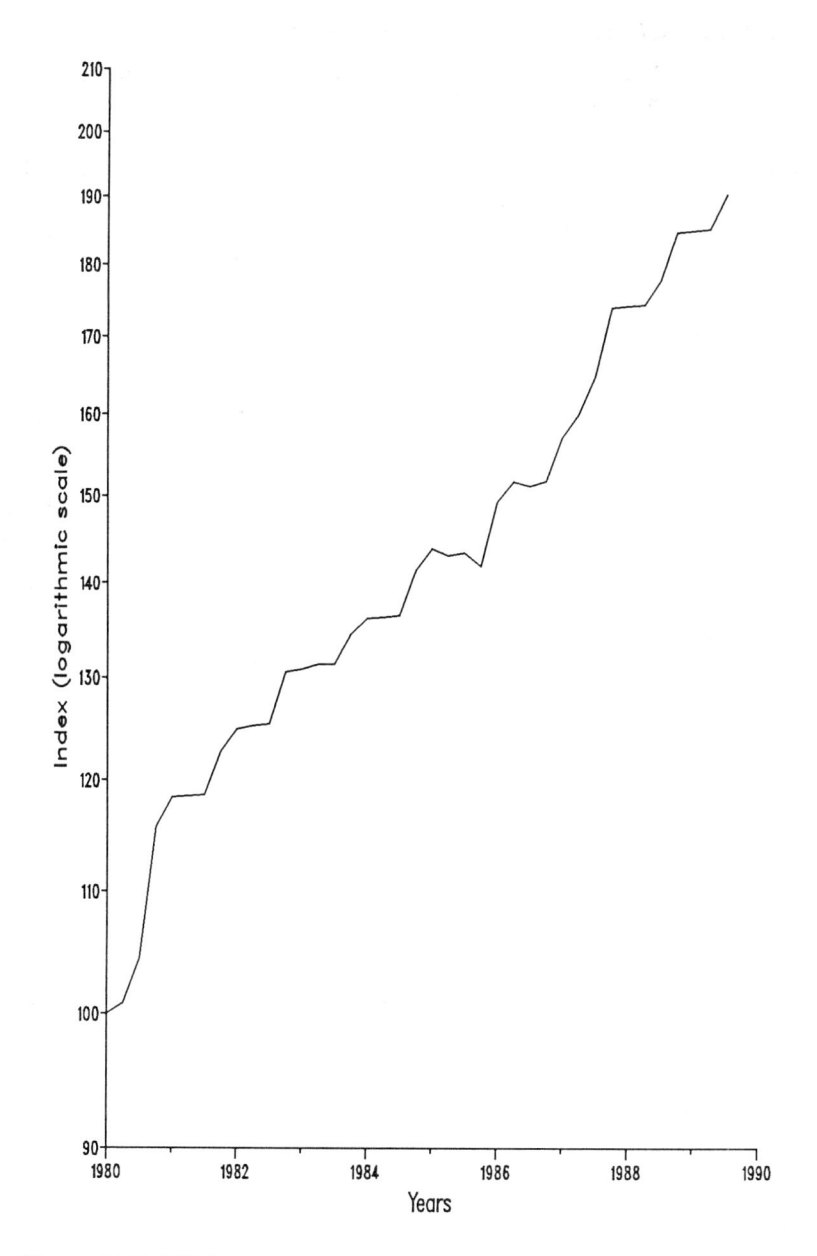

Figure 6.7.5 BMI Local Authority Maintenance Cost Index – Redecorations (1980 Q1 = 100), 1980 Q1 – 1989 Q3

Table 6.7.6 BMI Local Authority Maintenance Cost Index - Fabric Maintenance (BMI Table 2.2)

Base: 1980 Q1 = 100

Year	Q1	Q2	Q3	Q4	Average
1980	100.0	102.0	105.4	114.2	105.4
1981	116.4	117.7	118.2	121.6	118.5
1982	123.8	124.7	125.3	129.5	125.8
1983	130.1	131.6	132.2	135.2	132.3
1984	137.2	138.1	138.7	143.3	139.3
1985	146.3	146.0	146.4	146.6	146.3
1986	150.6	152.6	152.8	153.7	152.4
1987	158.7	161.2	165.2	172.3	164.4
1988	173.1	174.1	177.6	182.9	176.9
1989	184.0	184.8	189.2**		
1990					
1991					
1992					
1993					
1994					
1995					

Source: BMI.

** Provisional.

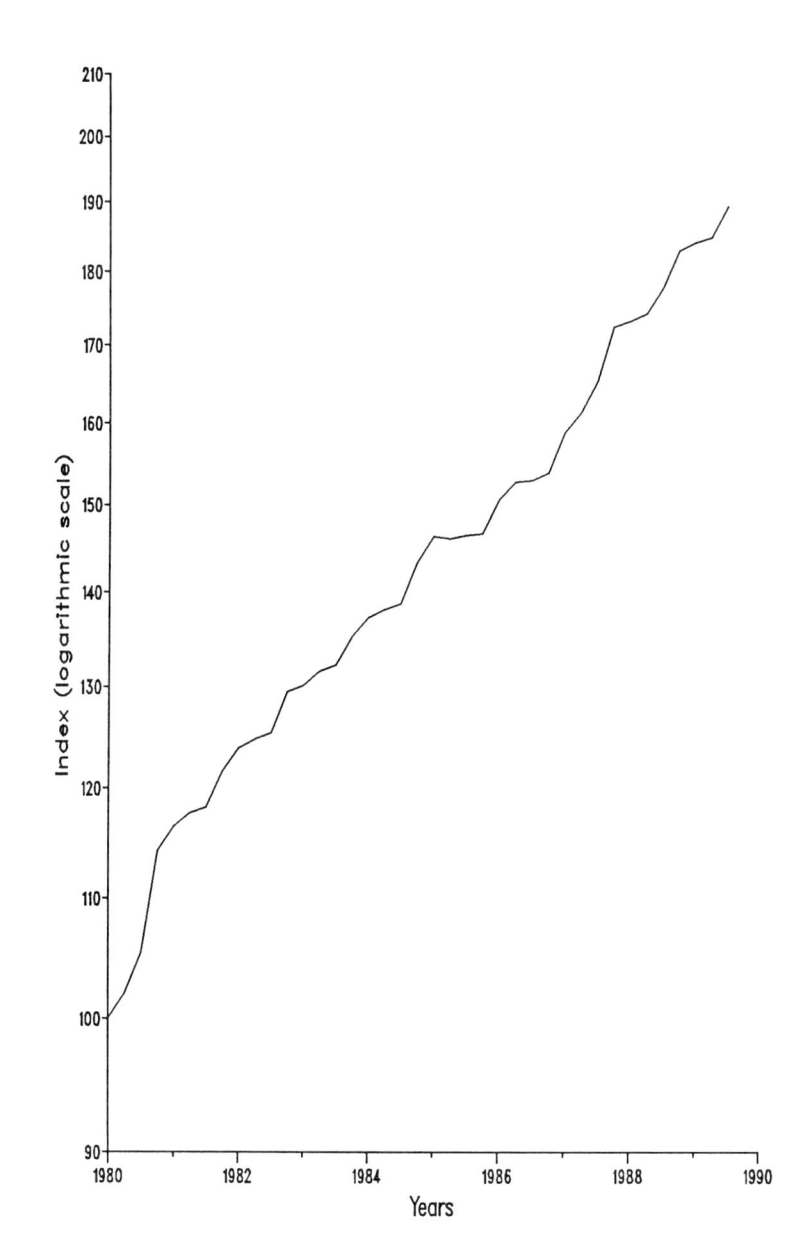

Figure 6.7.6 BMI Local Authority Maintenance Cost Index – Fabric Maintenance (1980 Q1 = 100), 1980 Q1 – 1989 Q3

Table 6.7.7 BMI Local Authority Maintenance Cost Index - Services Maintenance (BMI Table 2.3)

Base: 1980 Q1 = 100

Year	Q1	Q2	Q3	Q4	Average
1980	100.0	103.4	105.2	112.6	105.3
1981	114.8	115.4	116.0	119.7	116.5
1982	121.9	122.9	123.2	127.6	123.9
1983	128.4	129.5	129.6	132.1	129.9
1984	134.2	134.7	134.9	139.2	135.8
1985	142.4	141.3	141.1	140.8	141.4
1986	145.5	147.3	147.2	148.1	147.0
1987	152.3	154.4	157.8	163.6	157.0
1988	170.2	173.3	176.6	182.8	175.7
1989	184.1	184.5	188.9**		
1990					
1991					
1992					
1993					
1994					
1995					

Source: BMI.

** Provisional.

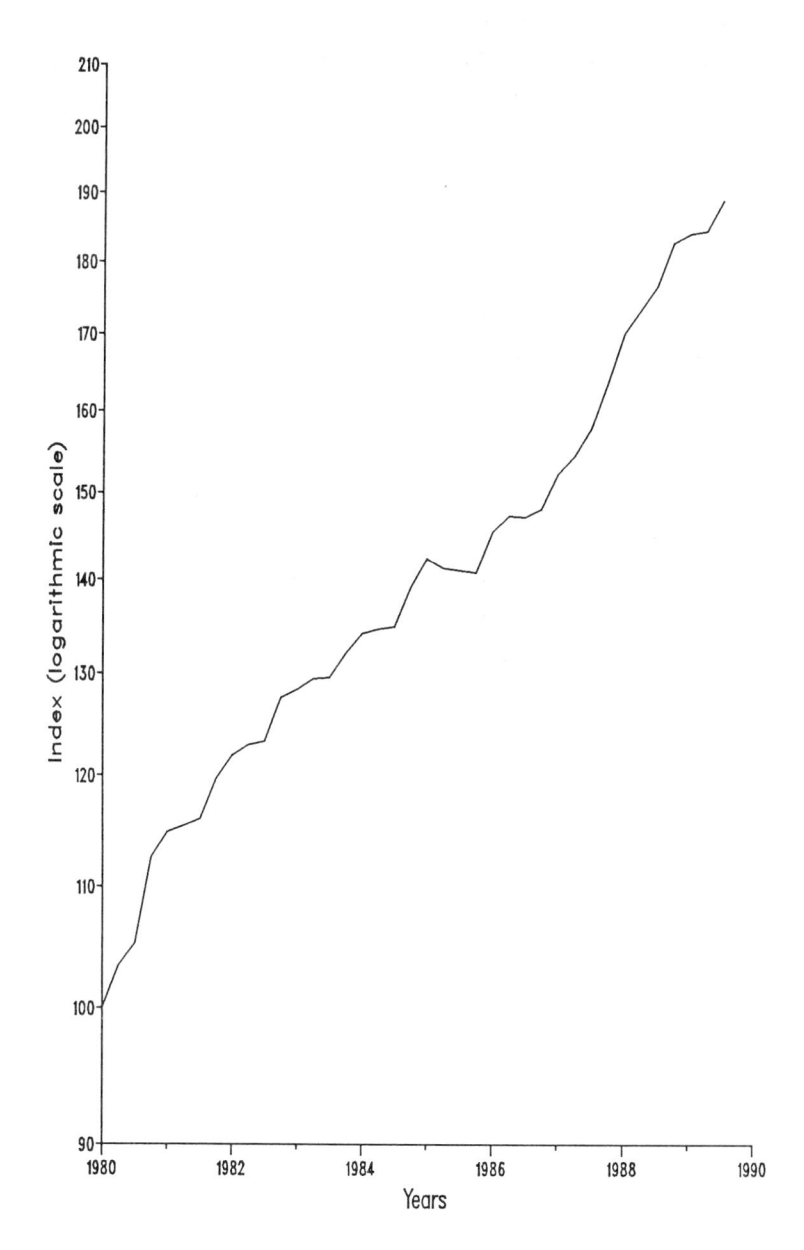

Figure 6.7.7. BMI Local Authority Maintenance Cost Index – Services Maintenance (1980 Q1 = 100), 1980 Q1 – 1989 Q3

Table 6.7.8 BMI Local Authority Maintenance Cost Index - General Maintenance (BMI Table 2.4)

Base: 1980 Q1 = 100

Year	Q1	Q2	Q3	Q4	Average
1980	100.0	102.2	105.0	114.1	105.3
1981	116.4	117.1	117.5	121.2	118.1
1982	123.4	124.2	124.5	129.1	125.3
1983	129.8	130.8	130.9	133.8	131.3
1984	135.7	136.3	136.6	140.9	137.4
1985	144.1	143.3	143.4	142.9	143.4
1986	148.2	150.3	150.2	151.0	149.9
1987	155.7	158.3	162.3	169.5	161.5
1988	172.3	173.9	177.1	183.5	176.7
1989	184.3	184.8	189.4**		
1990					
1991					
1992					
1993					
1994					
1995					

Source: BMI.

** Provisional.

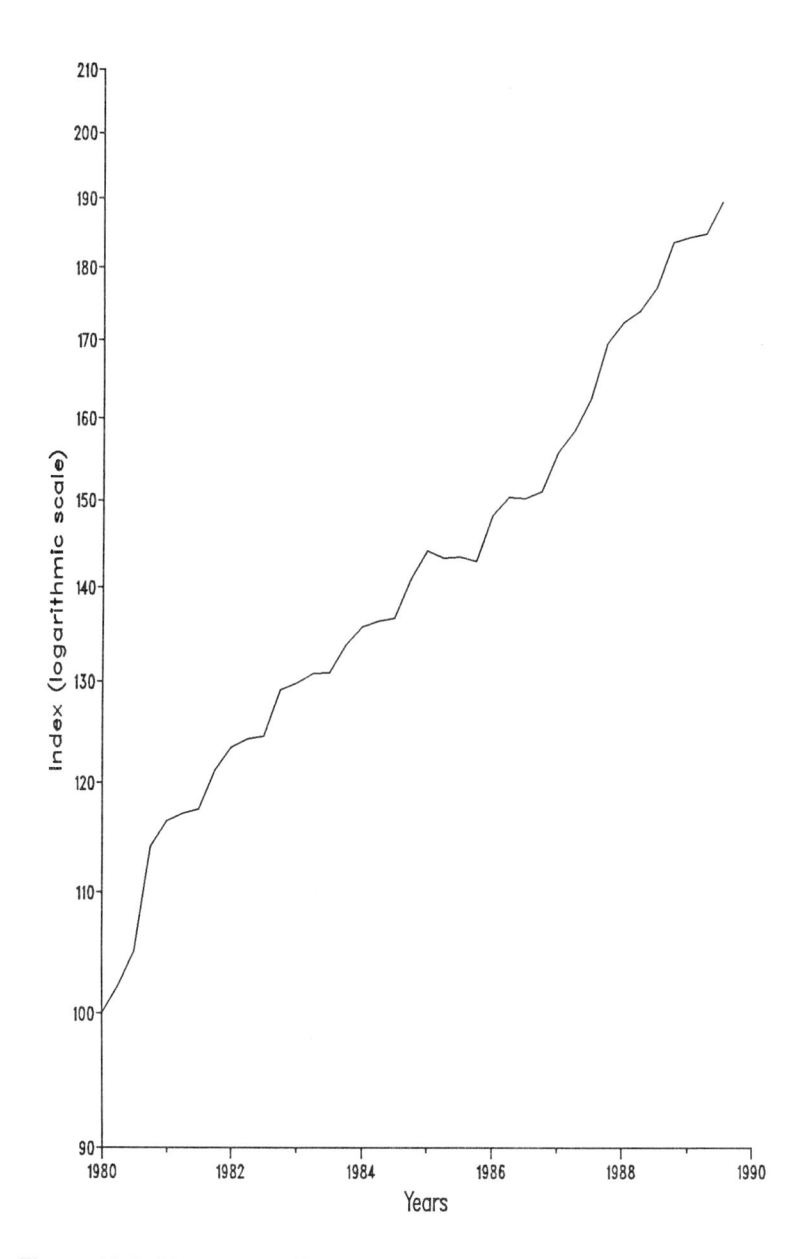

Figure 6.7.8 BMI Local Authority Maintenance Cost Index – General Maintenance (1980 Q1 = 100), 1980 Q1 – 1989 Q3

Table 6.7.9 BMI Private Sector Maintenance Cost Index - Redecorations (BMI Table 3.1)

Base: 1980 Q1 = 100

Year Average	Q1	Q2	Q3	Q4
1980	100.0	106.7	119.0	119.2	111.2
1981	119.4	119.7	127.2	129.7	124.0
1982	131.3	134.1	139.5	138.9	136.0
1983	138.7	139.1	146.0	146.2	142.5
1984	146.3	146.4	153.2	152.4	149.6
1985	152.5	152.8	160.1	160.5	156.5
1986	160.6	160.1	166.6	166.9	163.6
1987	167.2	167.6	175.2	175.3	171.3
1988	175.6	175.7	186.9	187.9	181.5
1989	188.0	188.4	205.4**		
1990					
1991					
1992					
1993					
1994					
1995					

Source: BMI.

** Provisional.

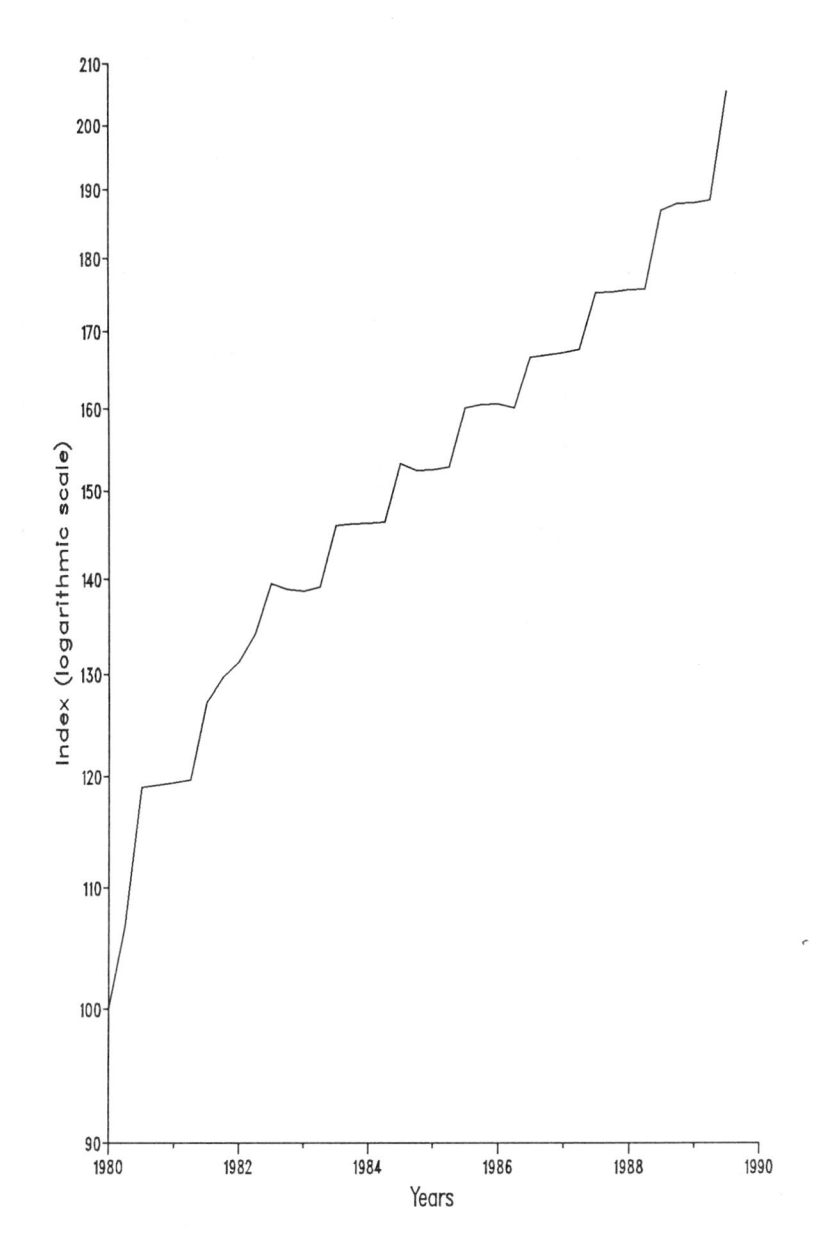

Figure 6.7.9 BMI Private Sector Maintenance Cost Index – Redecorations (1980 Q1 = 100), 1980 Q1 – 1989 Q3

Table 6.7.10 BMI Private Sector Maintenance Cost Index - Fabric Maintenance (BMI Table 3.2)

Base: 1980 Q1 = 100

Year	Q1	Q2	Q3	Q4	Average
1980	100.0	106.3	116.0	116.7	109.8
1981	117.1	118.5	124.5	126.8	121.7
1982	128.5	131.2	135.3	135.7	132.7
1983	136.3	137.2	142.9	143.7	140.0
1984	144.5	145.5	151.0	151.3	148.1
1985	152.6	153.1	158.6	159.0	155.8
1986	158.8	158.8	163.8	164.8	161.6
1987	166.1	166.8	172.8	173.3	169.8
1988	174.0	175.1	184.1	185.2	179.6
1989	186.3	187.1	200.3**		
1990					
1991					
1992					
1993					
1994					
1995					

Source: BMI.

** Provisional.

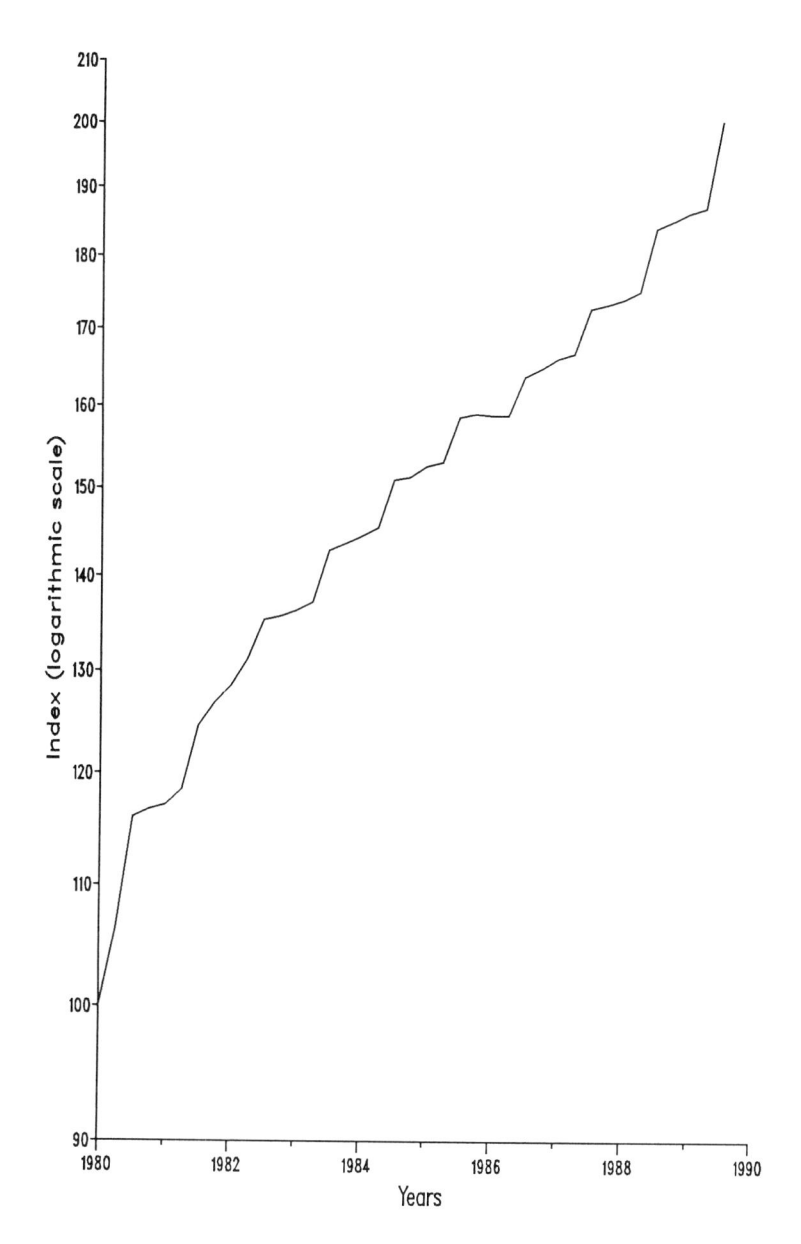

Figure 6.7.10 BMI Private Sector Maintenance Cost Index – Fabric Maintenance (1980 Q1 = 100), 1980 Q1 – 1989 Q3

Table 6.7.11 BMI Private Sector Maintenance Cost Index - Services Maintenance (BMI Table 3.3)

Base: 1980 Q1 = 100

Year	Q1	Q2	Q3	Q4	Average
1980	100.0	102.1	104.1	108.0	103.6
1981	113.7	116.7	117.9	118.8	116.8
1982	121.1	128.0	127.5	127.9	126.1
1983	129.5	135.0	136.6	136.9	134.5
1984	137.7	141.3	143.5	143.5	141.5
1985	144.6	147.7	148.4	150.8	147.9
1986	151.4	153.3	153.4	156.4	153.6
1987	160.8	164.1	165.3	167.1	164.3
1988	171.2	175.6	176.5	178.3	175.4
1989	182.0	187.3	188.0**		
1990					
1991					
1992					
1993					
1994					
1995					

Source: BMI.

** Provisional.

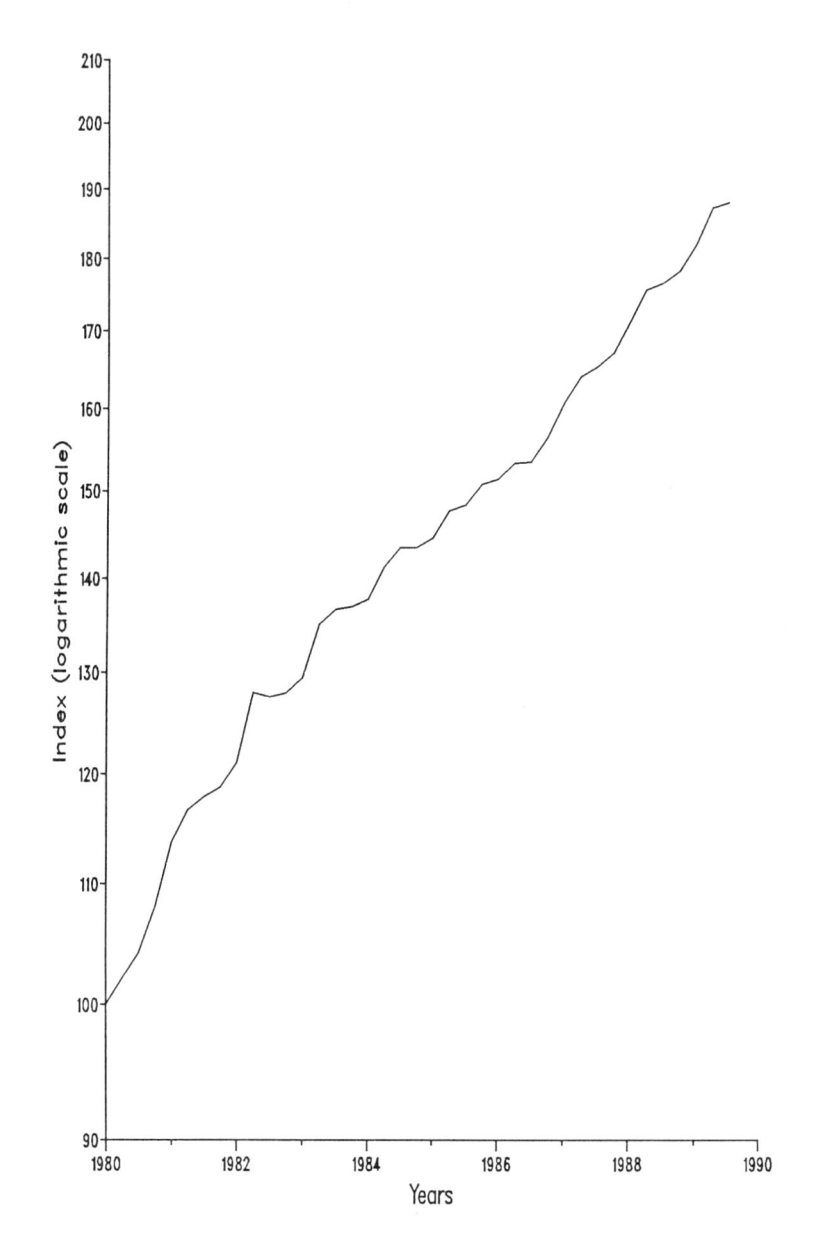

Figure 6.7.11 BMI Private Sector Maintenance Cost Index – Services Maintenance (1980 Q1 = 100), 1980 Q1 – 1989 Q3

Table 6.7.12 BMI Private Sector Maintenance Cost Index - General Maintenance (BMI Table 3.4)

Base: 1980 Q1 = 100

Year	Q1	Q2	Q3	Q4	Average
1980	100.0	105.0	113.4	114.2	108.2
1981	116.5	118.2	122.8	124.4	120.5
1982	126.6	130.9	133.6	133.8	131.2
1983	134.5	137.0	141.5	141.9	138.7
1984	142.5	144.2	148.8	148.7	146.1
1985	149.5	150.9	155.2	156.4	153.0
1986	156.6	157.1	160.7	162.3	159.2
1987	164.4	166.0	170.7	171.6	168.2
1988	173.5	175.5	182.1	183.5	178.7
1989	185.2	187.6	197.3**		
1990					
1991					
1992					
1993					
1994					
1995					

Source: BMI.

** Provisional.

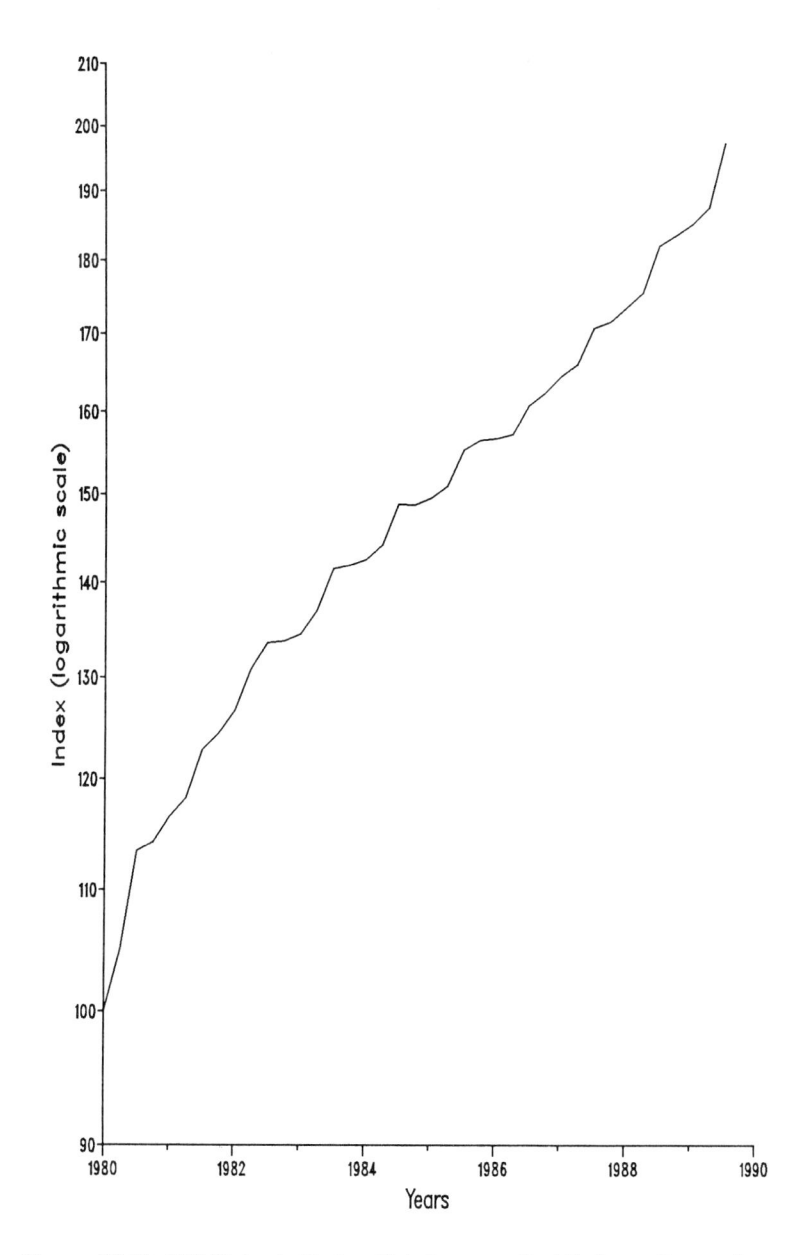

Figure 6.7.12 BMI Private Sector Maintenance Cost Index – General Maintenance (1980 Q1 = 100), 1980 Q1 – 1989 Q3

Table 6.7.13 BMI All-In Maintenance Cost Index - Redecorations (BMI Table 4.1)

Base: 1980 Q1 = 100

Year	Q1	Q2	Q3	Q4	Average
1980	100.0	102.7	108.2	112.1	105.8
1981	120.7	120.9	123.4	125.7	122.7
1982	129.7	130.8	132.2	133.8	131.6
1983	136.9	137.3	139.5	140.7	138.6
1984	141.3	143.5	145.9	147.2	144.5
1985	148.2	149.8	152.5	151.6	150.5
1986	154.1	157.8	159.8	160.3	158.0
1987	162.3	169.8	174.1	177.2	170.9
1988	177.6	180.7	185.7	188.7	183.2
1989	188.8	193.0	200.4**		
1990					
1991					
1992					
1993					
1994					
1995					

Source: BMI.

** Provisional.

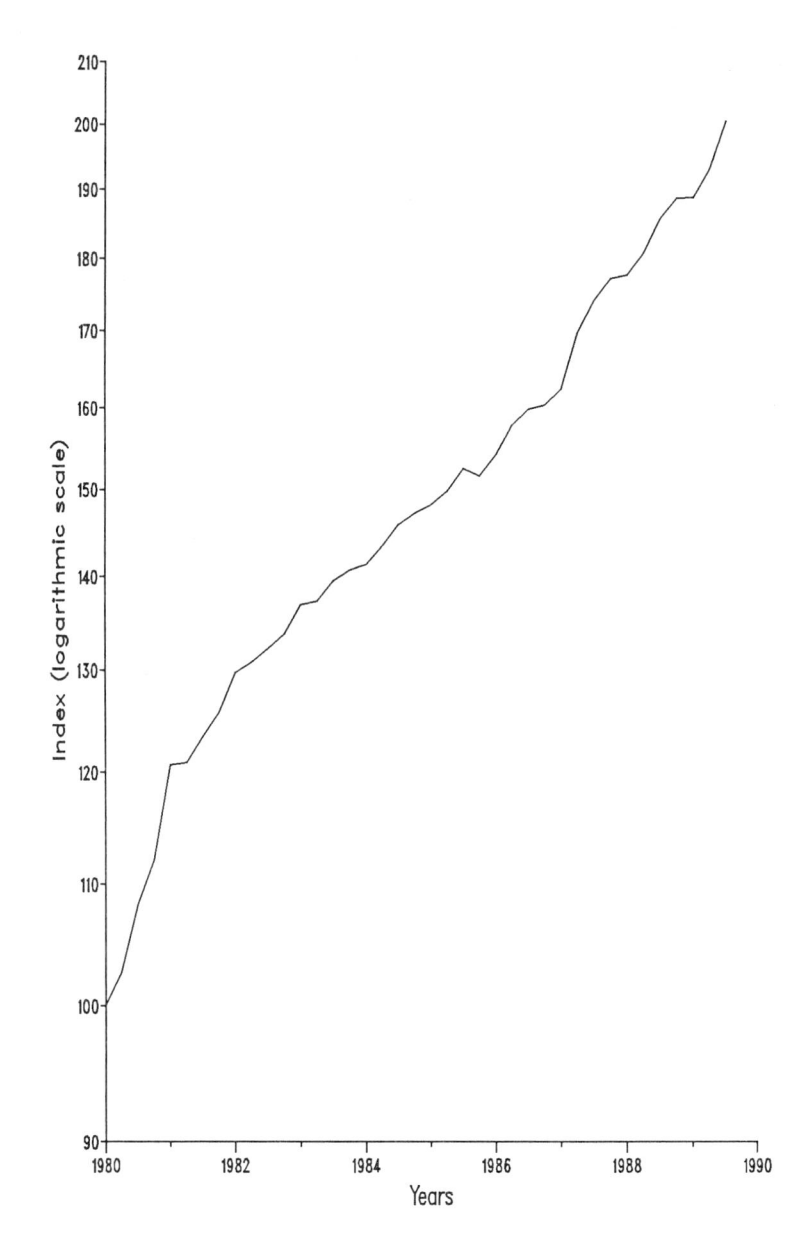

Figure 6.7.13 BMI All-in Maintenance Cost Index – Redecorations (1980 Q1 = 100), 1980 Q1 – 1989 Q3

Table 6.7.14 BMI All-In Maintenance Cost Index - Fabric Maintenance (BMI Table 4.2)

Base: 1980 Q1 = 100

Year	Q1	Q2	Q3	Q4	Average
1980	100.0	103.5	108.1	111.5	105.8
1981	118.0	119.4	121.7	123.8	120.7
1982	127.3	128.8	130.3	131.9	129.6
1983	134.8	135.9	138.2	139.7	137.2
1984	140.9	143.4	145.6	147.5	144.4
1985	149.5	151.0	153.1	152.8	151.6
1986	154.1	157.1	159.1	159.9	157.6
1987	162.4	168.4	172.0	174.7	169.4
1988	175.5	178.7	183.3	185.8	180.9
1989	186.8	190.4	196.6**		
1990					
1991					
1992					
1993					
1994					
1995					

Source: BMI.

** Provisional.

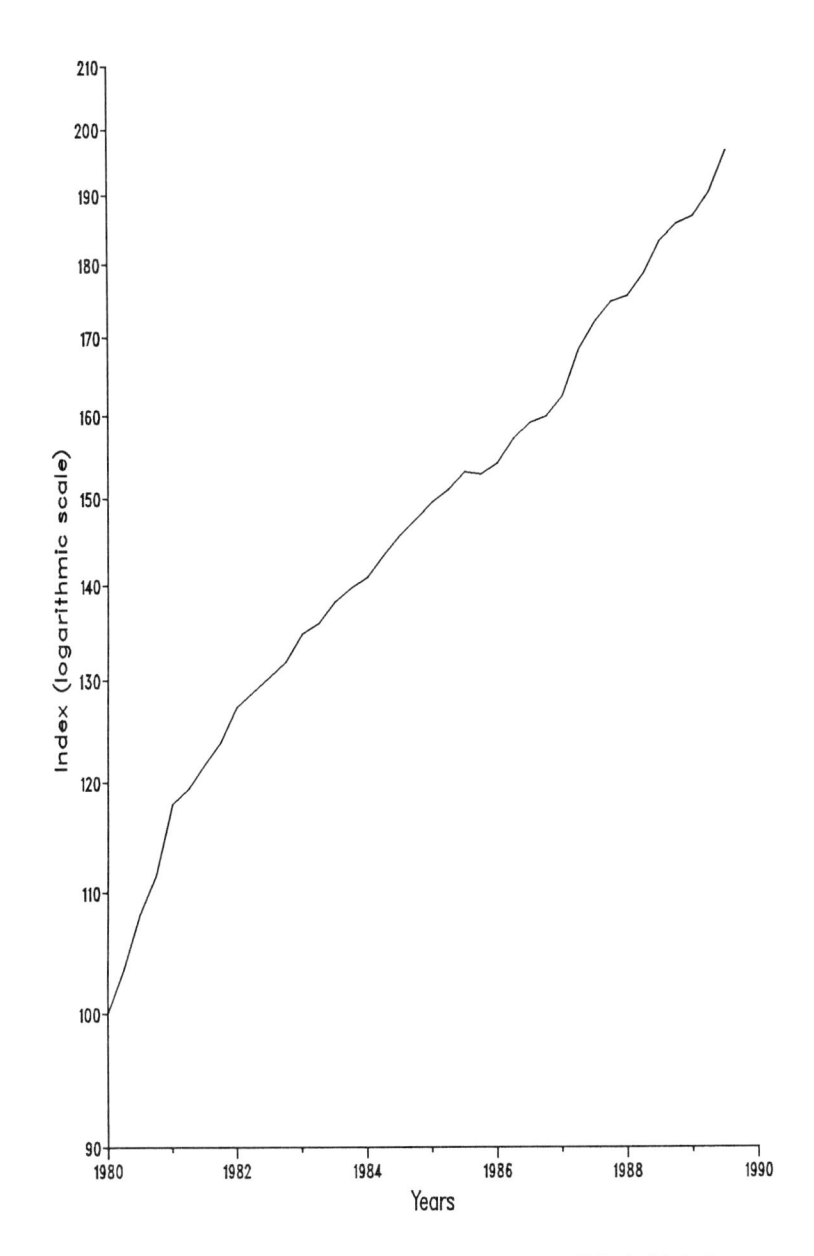

Figure 6.7.14 BMI All-in Maintenance Cost Index – Fabric Maintenance
(1980 Q1 = 100), 1980 Q1 – 1989 Q3

Table 6.7.15 BMI All-In Maintenance Cost Index - Services Maintenance (BMI Table 4.3)

Base: 1980 Q1 = 100

Year	Q1	Q2	Q3	Q4	Average
1980	100.0	102.2	103.5	107.5	103.3
1981	115.8	117.2	117.9	119.8	117.7
1982	123.6	126.5	126.3	127.9	126.1
1983	131.8	134.4	135.0	135.9	134.3
1984	137.2	140.7	141.5	142.9	140.6
1985	144.9	145.5	147.0	148.1	146.4
1986	149.7	151.4	152.8	155.2	152.3
1987	162.1	166.0	169.0	172.3	167.4
1988	176.6	179.3	182.6	186.5	181.3
1989	188.7	190.6	195.4**		
1990					
1991					
1992					
1993					
1994					
1995					

Source: BMI.

** Provisional.

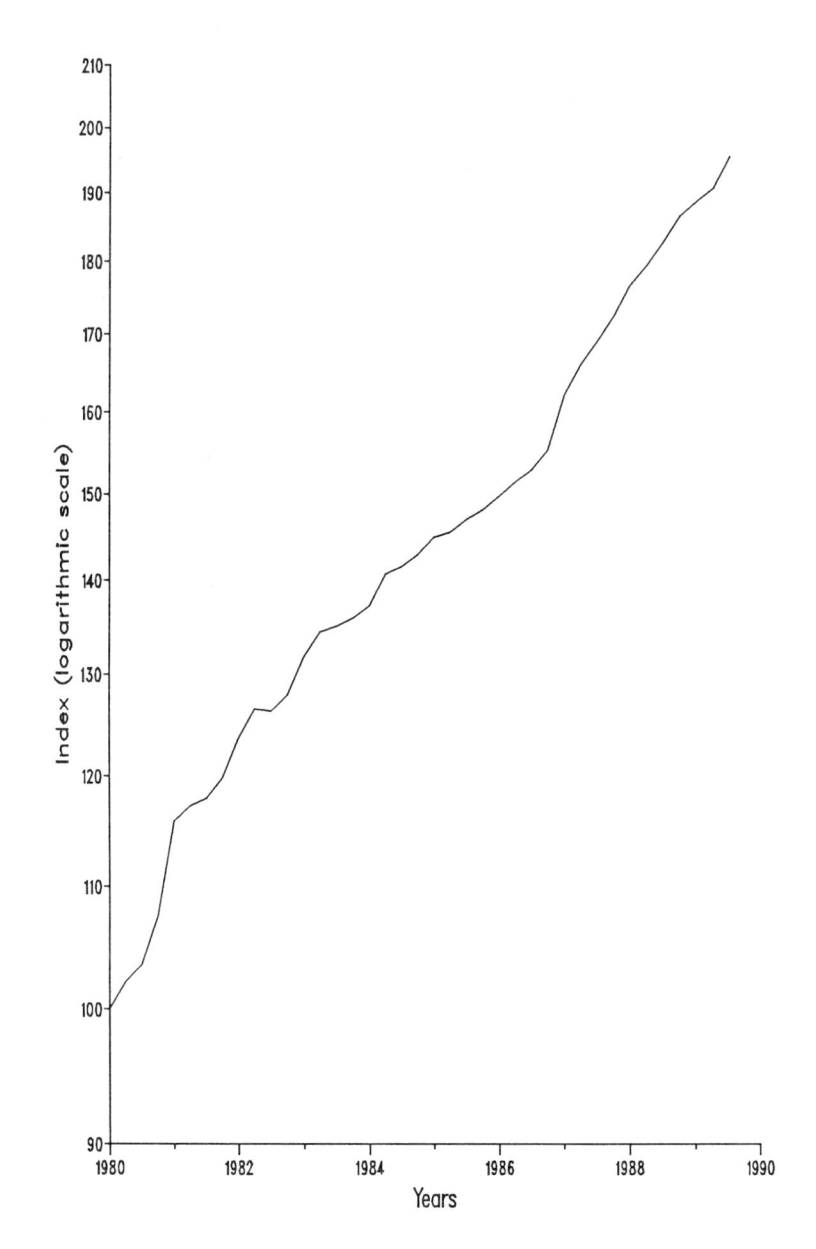

Figure 6.7.15 BMI All-in Maintenance Cost Index − Services Maintenance (1980 Q1 = 100), 1980 Q1 − 1989 Q3

Table 6.7.16 BMI All-In Maintenance Cost Index - General Maintenance (BMI Table 4.4)

Base: 1980 Q1 = 100

Year	Q1	Q2	Q3	Q4	Average
1980	100.0	102.7	106.4	110.2	104.8
1981	118.0	119.0	120.8	122.9	120.2
1982	126.6	128.6	129.4	131.0	128.9
1983	134.4	135.8	137.2	138.6	136.5
1984	139.6	142.4	144.1	145.7	143.0
1985	147.3	148.5	150.4	150.6	149.2
1986	152.4	154.8	156.9	158.2	155.6
1987	162.0	167.9	171.5	174.6	169.0
1988	176.6	179.6	183.8	187.0	181.8
1989	188.2	191.3	197.3**		
1990					
1991					
1992					
1993					
1994					
1995					

Source: BMI.

** Provisional.

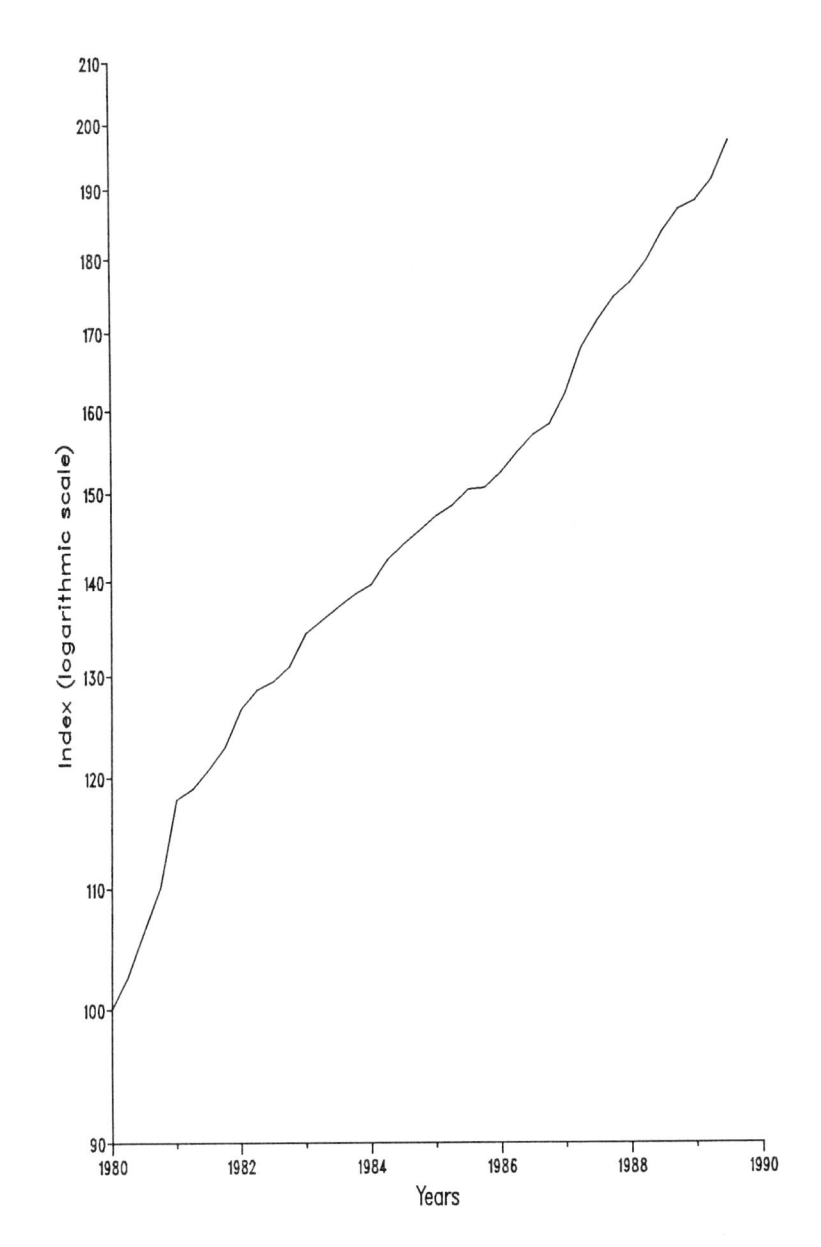

Figure 6.7.16 BMI All-in Maintenance Cost Index – General Maintenance (1980 Q1 = 100), 1980 Q1 – 1989 Q3

Table 6.7.17 BMI Maintenance Materials Cost Index - Redecorations (BMI Table 5.1)

Base: 1980 Q1 = 100

Year	Q1	Q2	Q3	Q4	Average
1980	100.0	105.4	109.9	110.0	106.3
1981	112.2	113.1	113.9	115.1	113.6
1982	117.4	120.0	122.1	122.3	120.5
1983	122.4	124.0	126.3	128.3	125.3
1984	128.9	130.3	131.7	133.3	131.1
1985	134.5	137.0	140.3	142.2	138.5
1986	142.5	138.5	136.6	139.8	139.4
1987	142.3	145.3	148.5	149.4	146.4
1988	152.2	153.2	157.0	164.6	156.8
1989	165.4	169.4	171.9**		
1990					
1991					
1992					
1993					
1994					
1995					

Source: BMI.

** Provisional.

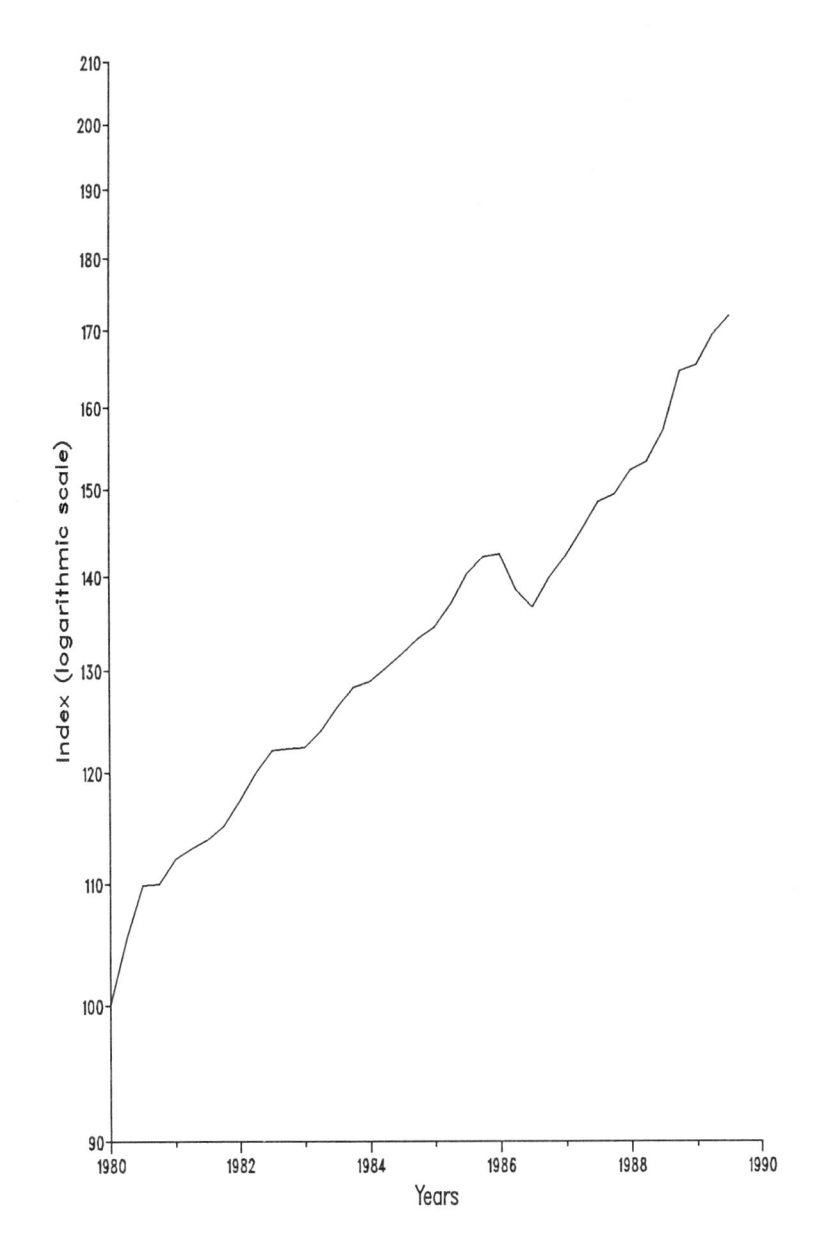

Figure 6.7.17 BMI Maintenance Materials Cost Index – Redecorations (1980 Q1 = 100), 1980 Q1 – 1989 Q3

Table 6.7.18 BMI Maintenance Materials Cost Index - Fabric Maintenance (BMI Table 5.2)

Base: 1980 Q1 = 100

Year	Q1	Q2	Q3	Q4	Average
1980	100.0	105.5	108.4	110.0	106.0
1981	111.1	115.0	116.4	118.0	115.1
1982	120.0	122.6	124.4	125.7	123.2
1983	127.2	130.1	132.5	135.0	131.2
1984	137.2	140.0	141.8	144.8	141.0
1985	148.6	150.2	151.3	151.8	150.5
1986	151.4	151.3	151.8	154.4	152.2
1987	158.2	160.4	162.3	163.6	161.1
1988	166.0	168.7	172.5	174.7	170.5
1989	177.8	180.2	182.9**		
1990					
1991					
1992					
1993					
1994					
1995					

Source: BMI.

** Provisional.

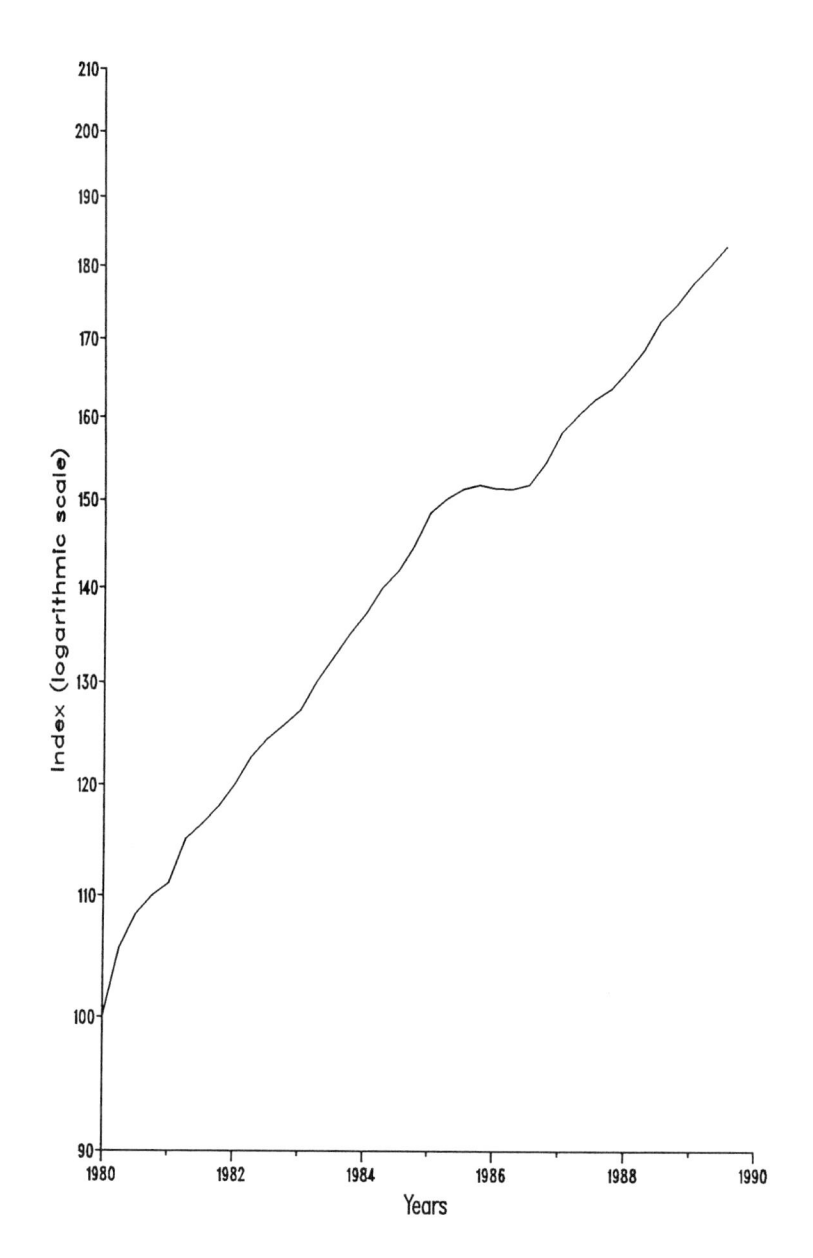

Figure 6.7.18 BMI Maintenance Materials Cost Index – Fabric Maintenance (1980 Q1 = 100), 1980 Q1 – 1989 Q3

Table 6.7.19 BMI Maintenance Materials Cost Index - Services Maintenance (BMI Table 5.3)

Base: 1980 Q1 = 100

Year	Q1	Q2	Q3	Q4	Average
1980	100.0	102.6	103.3	104.8	102.7
1981	105.7	107.4	109.3	112.0	108.6
1982	114.7	117.5	118.3	119.9	117.6
1983	121.2	123.6	124.5	125.0	123.6
1984	127.6	129.1	129.6	131.8	129.5
1985	135.9	135.3	134.5	134.9	135.1
1986	136.6	136.4	136.2	138.6	137.0
1987	140.2	141.0	144.5	148.7	143.6
1988	152.7	154.1	157.0	161.3	156.3
1989	165.0	166.3	188.3**		
1990					
1991					
1992					
1993					
1994					
1995					

Source: BMI.

** Provisional.

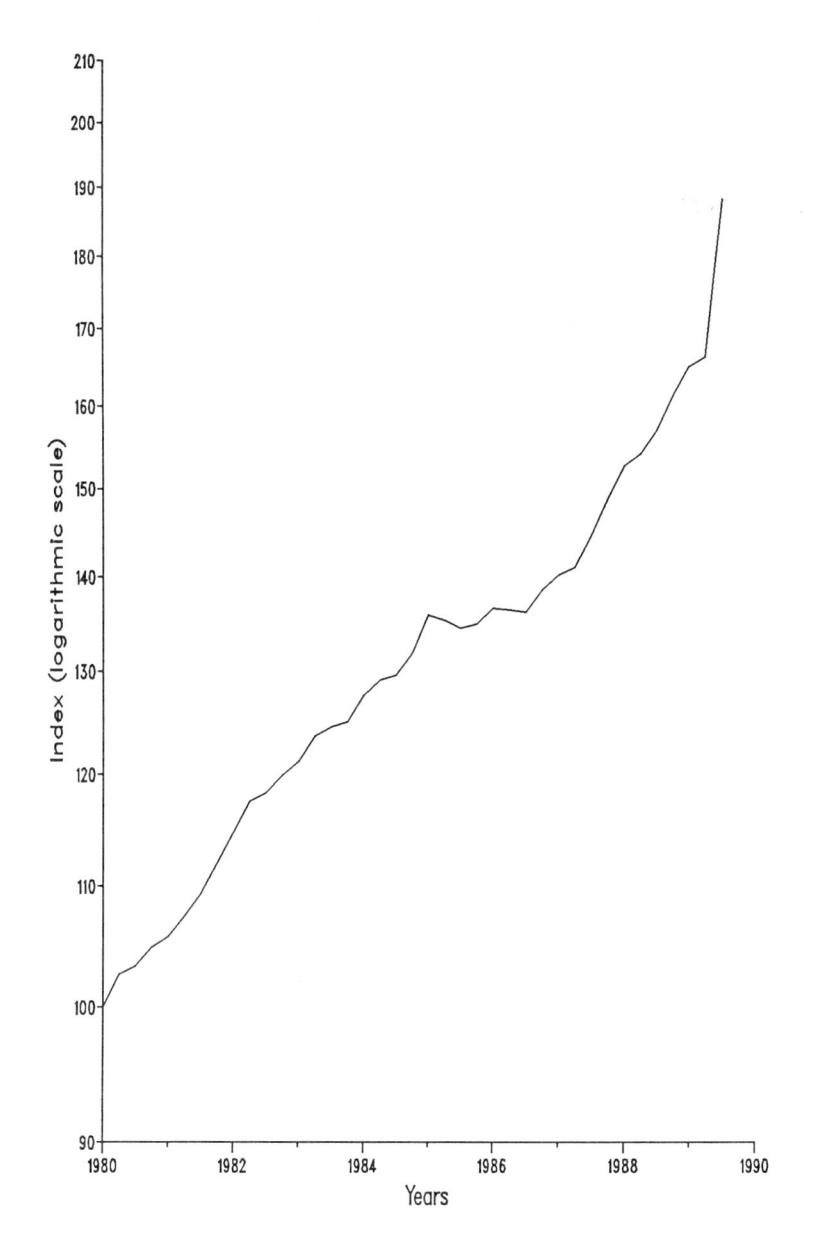

Figure 6.7.19 BMI Maintenance Materials Cost Index − Services Maintenance (1980 Q1 = 100), 1980 Q1 − 1989 Q3

Table 6.7.20 BMI Maintenance Materials Cost Index - General Maintenance (BMI Table 5.4)

Base: 1980 Q1 = 100

Year	Q1	Q2	Q3	Q4	Average
1980	100.0	104.1	106.1	107.4	104.4
1981	108.6	111.1	112.6	114.7	111.8
1982	117.1	119.7	121.1	122.4	120.1
1983	123.7	126.1	127.8	129.3	126.7
1984	131.4	133.4	134.5	137.0	134.1
1985	140.6	141.2	141.6	142.3	141.4
1986	143.0	142.4	142.2	144.8	143.1
1987	147.3	148.9	151.8	154.4	150.6
1988	157.6	159.5	162.4	166.2	160.9
1989	170.0	172.0	174.3**		
1990					
1991					
1992					
1993					
1994					
1995					

Source: BMI.

** Provisional.

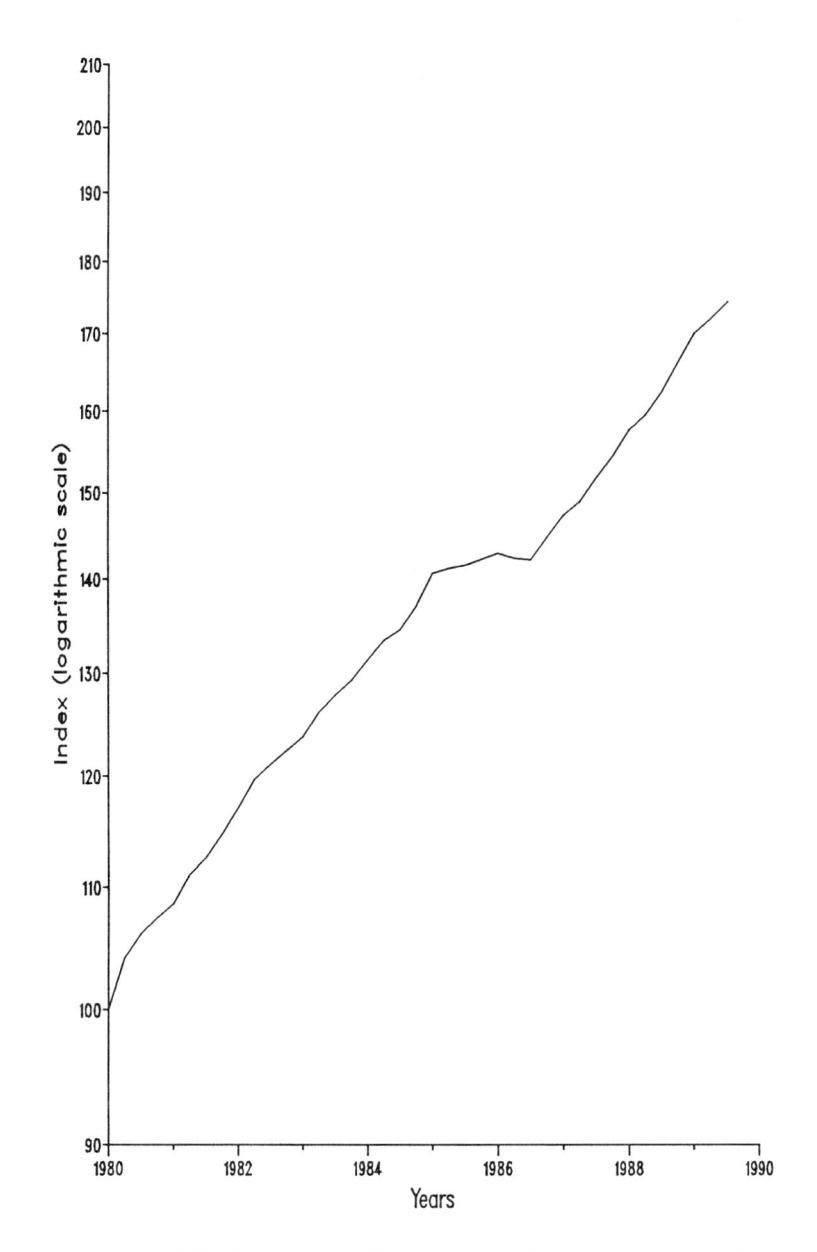

Figure 6.7.20 BMI Maintenance Materials Cost Index – General Maintenance (1980 Q1 = 100), 1980 Q1 – 1989 Q3

Table 6.7.21 BMI Cleaning Costs Index (BMI Table 7.1)

Base: 1980 Q1 = 100

Year	Q1	Q2	Q3	Q4	Average
1980	100.0	103.4	103.5	109.5	104.1
1981	112.0	112.1	112.2	117.5	113.5
1982	120.0	120.2	119.8	125.2	121.3
1983	127.7	128.1	128.1	132.0	129.0
1984	134.3	134.4	134.6	139.5	135.7
1985	141.4	141.4	146.8	157.3	146.7
1986	157.5	157.6	159.9	164.4	159.9
1987	164.6	164.7	175.6	175.7	170.2
1988	176.0	176.4	179.5	186.0	179.5
1989	186.1	186.2	191.6**		
1990					
1991					
1992					
1993					
1994					
1995					

Source: BMI.

** Provisional.

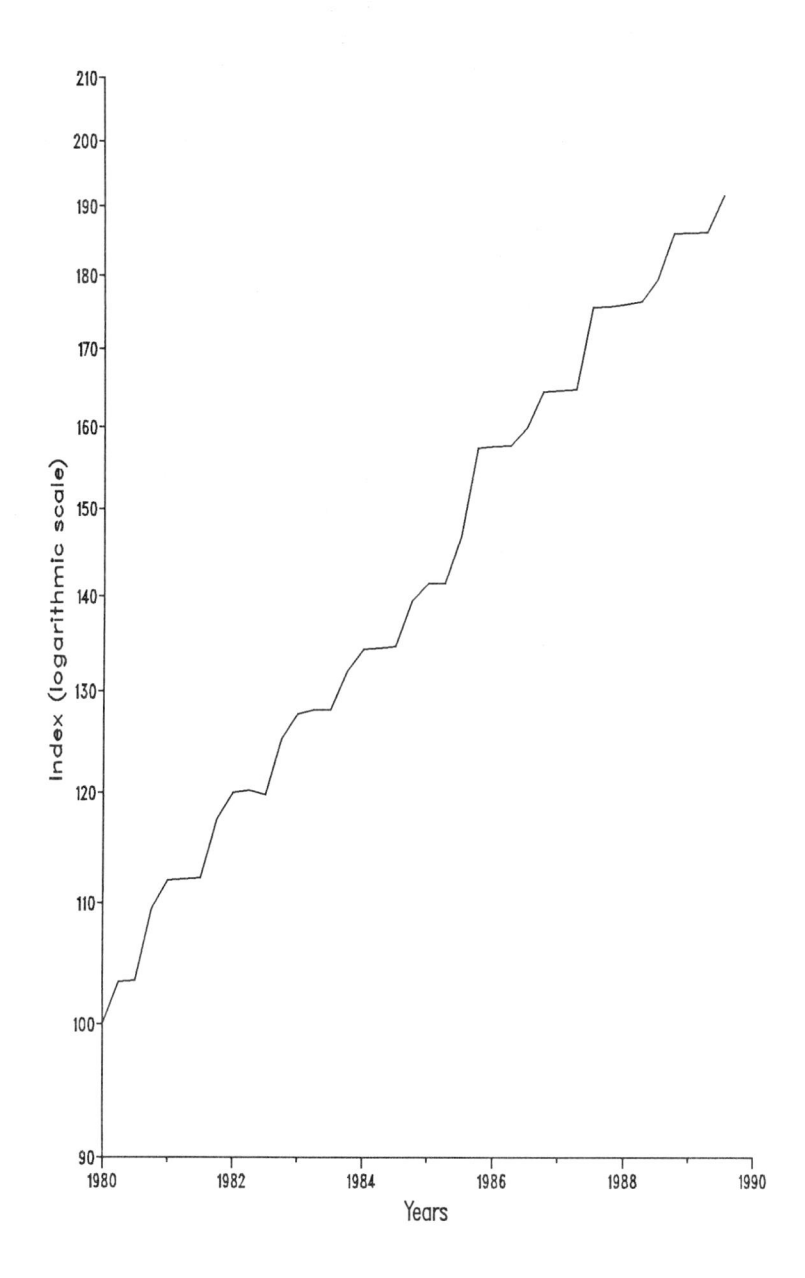

Figure 6.7.21 BMI Cleaning Costs Index (1980 Q1 = 100), 1980 Q1 – 1989 Q3

Table 6.7.22 BMI Cleaning Materials Costs Index (BMI Table 7.2)

Base: 1980 Q1 = 100

Year	Q1	Q2	Q3	Q4	Average
1980	100.0	104.4	107.3	108.3	105.0
1981	109.2	109.9	110.5	112.0	110.4
1982	114.9	117.4	119.4	121.2	118.2
1983	122.7	124.0	126.7	128.0	125.4
1984	131.9	133.1	135.5	138.4	134.7
1985	140.4	142.5	145.2	145.8	143.5
1986	148.1	148.8	148.9	149.3	148.8
1987	151.4	153.3	154.6	156.7	154.0
1988	159.9	160.1	164.8	166.4	162.8
1989	167.7	169.2	172.6**		
1990					
1991					
1992					
1993					
1994					
1995					

Source: BMI.

** Provisional.

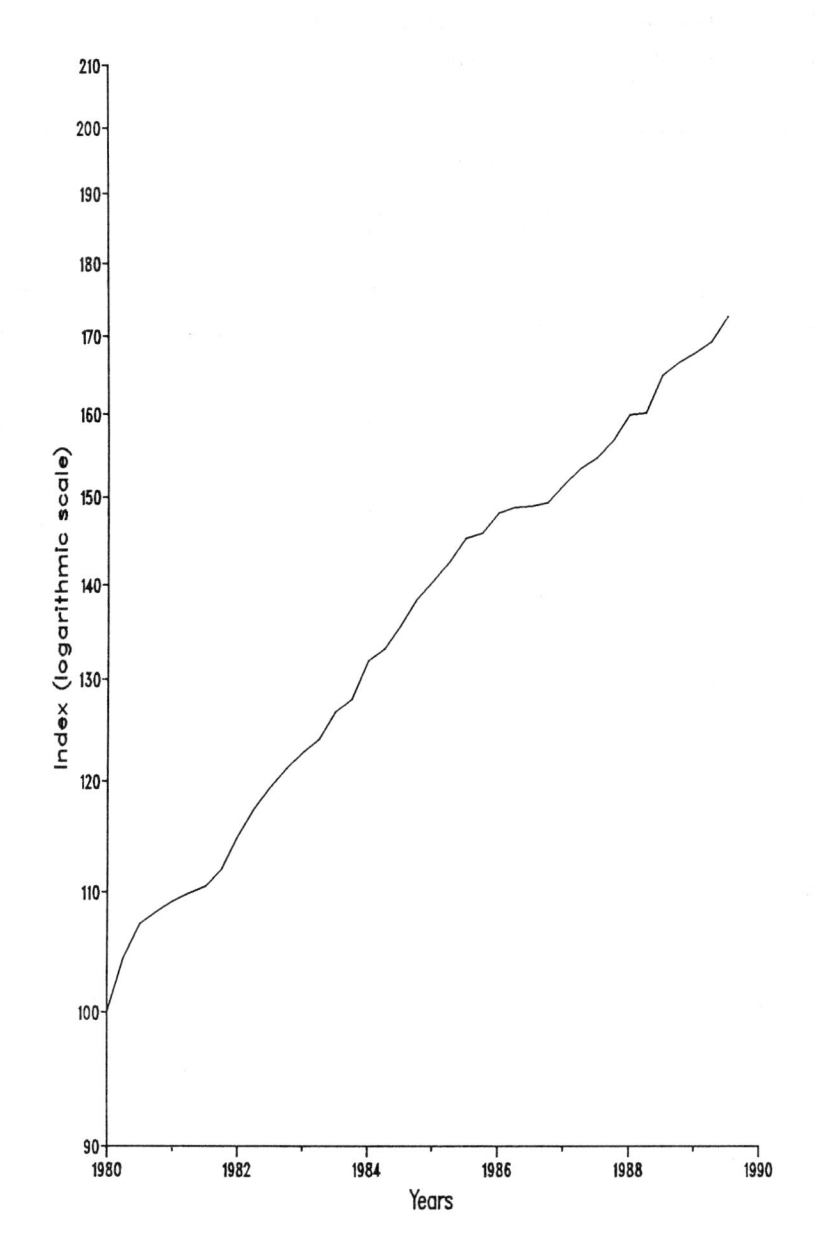

Figure 6.7.22 BMI Cleaning Materials Costs Index (1980 = 100), 1980 Q1 – 1989 Q3

Table 6.7.23 BMI Energy Cost Index (BMI Table 8.1)

Base: 1980 Q1 = 100

Year	Q1	Q2	Q3	Q4	Average
1980	100	105	114	117	109
1981	121	125	130	136	128
1982	137	137	143	153	143
1983	156	149	151	153	152
1984	156	152	154	166	157
1985	174	161	163	166	167
1986	160	148	141	153	150
1987	151	146	151	156	151
1988	156**	154**	156	162**	157**
1989	160**	162**			
1990					
1991					
1992					
1993					
1994					
1995					

Source: BMI.

** Provisional.

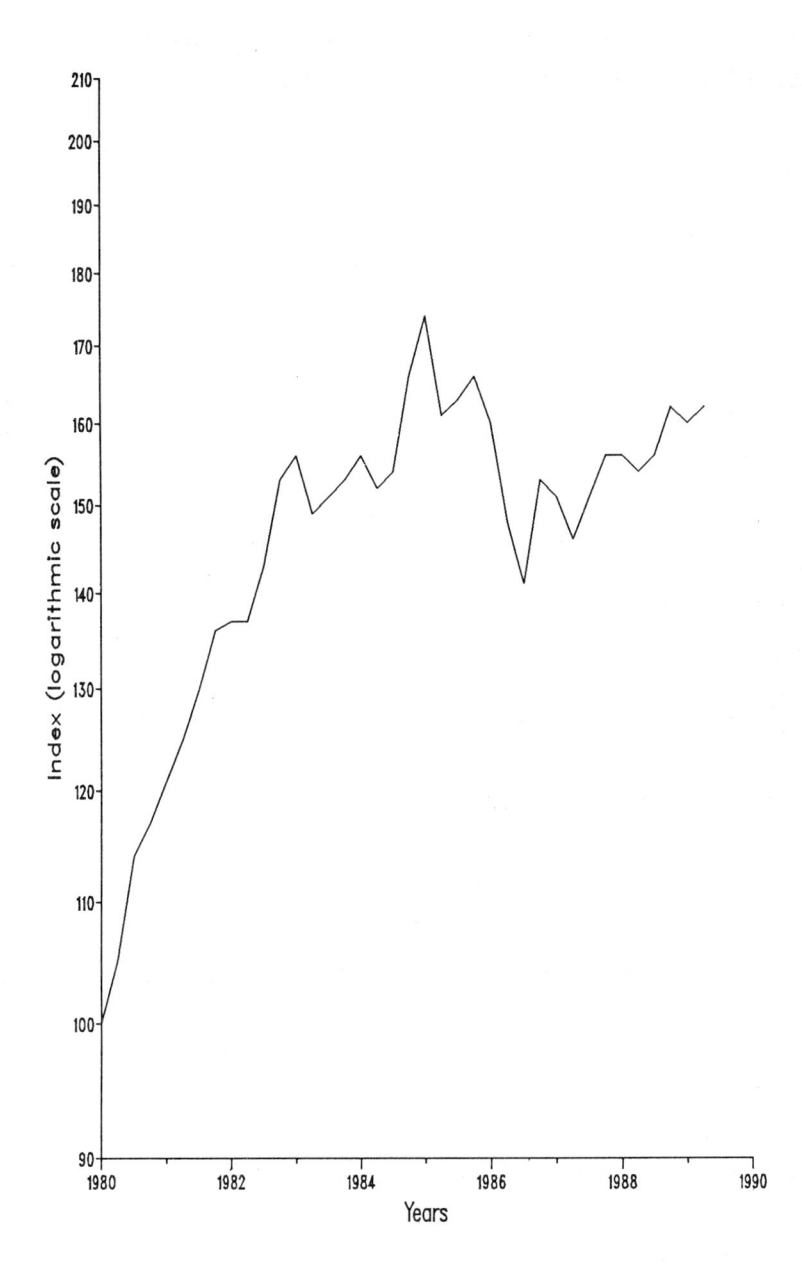

Figure 6.7.23 BMI Energy Cost Index (1980 Q1 = 100), 1980 Q1 – 1989 Q3

6.8 APSAB COST INDICES

Type of Index

Building cost index (BCI).

Series: Coverage and Breakdowns

Analyses of a new hypothetical building in the public or private sector with breakdowns as follows:

Series	Table Reference
Movement of building costs by application of the Price Adjustment Formula for Construction Contracts (APSAB index)	
Building, 1970 to date	6.8.1
H & V and air conditioning, 1975 to date	6.8.2
Electrical, 1975 to date	6.8.3
Combined index, 1975 to date	6.8.4

Base Dates and Period Covered

1970	from 1970 to 1977
1975	from 1975 to 1988
1985	from 1984 to date

The pre-1985 series are shown with 1985 = 100 in the tables below (converted arithmetically by the PSA).

Frequency

The formula figures are produced monthly and the APSAB indices are calculated monthly, but the APSAB indices are published as a quarterly series.

Geographical Coverage

United Kingdom.

Type and Source of Data

The inputs to the APSAB cost indices are the Work Category Indices (Series 2) prepared by the Property Services Agency for use with the Price Adjustment Formulae for Construction Contracts.

Method of Compilation

The formula method of price adjustment for building contracts is a method of calculating the reimbursement to or from a contractor of the increases or decreases in cost that occur during a building contract. The conventional method of calculating these increases or decreases is by the inspection of time and wage sheets, suppliers' invoices and the like. The formula method makes use of indices for this calculation to save time and effort.

The method employs indices for 49 (series 2) work categories for building broadly based on the traditional work sections.

The PSA, who produce the indices, analysed a large number of bills of quantities and arrived at typical proportions of labour, materials and plant in each category. From the individual indices for each of these three components a composite index for each category is compiled. These indices are based on market prices, nationally agreed wages etc. and do not take into account that it may be necessary for a contractor to pay for labour above these rates or buy materials at much higher or lower prices. Any payments which are 'at a premium or discount' are, therefore, not included at all, and the indices for each of the work categories is in essence a factor cost index. The indices for each work category are produced in *Price Adjustment Formulae for Construction Contracts: Monthly Bulletin Indices* for series 2, series 1 having been discontinued.

The formula method operates by having all the items in a bill of quantities allocated either into a work category, the balance of adjustable work (to be treated in the same proportion to the increase or decrease on the work attributable to the work categories) or specifically excluded (e.g. pc sums) to be adjusted separately by special formula. Bills of quantities treated in the foregoing manner are priced at rates current during the tender period and do not take into account any promulgated increases or decreases. The base month is stated in the bill of quantities.

The price fluctuations are calculated at each valuation and to do this the valuation has to be prepared with the amount of work done allocated to the relevant work categories as stated in the bill of quantities. Once the allocation has been made the calculation of the increase or decrease is arithmetically simple using the following formula:

$$C = \frac{V(I_v - I_0)}{I_0}$$

where C = the amount of the price adjustment for the work category to be paid to or recovered from the contractor.

V = the value of work executed in the work category during the valuation period.

I_V = the work category index number current at the mid-point of the valuation period.

I_0 = the work category index number for the base month.

An example is as follows:

	Index
Base month (I_0)	110
Valuation month (I_V)	121
Valuation for category (V)	£10,000

$$C = £10,000 \times \frac{121 - 110}{110} = £1,000$$

This figure may need to be subsequently adjusted to take into account any non-adjustable element necessary. Full details of the working of the formula method for building and specialist installations, which in itself is a very complex subject can be found in the many documents used in complying with it.

The indices published for use with the formula method have been put to another use by the PSA. The following is extracted from PSA *Quantity Surveyors Information Notes*:

'The APSAB index, compiled monthly, measures the movement in cost to contractors of construction of a typical PSA building. The index gives a guide to the average movement of building costs and has been used, in cost reimbursement contracts, for updating the estimated prime cost budget for comparison with actual prime costs. When compared with the Building Tender Price Index it can give some guidance to the influence of building costs on movement of tender prices; it does not contain any allowance for market factors, "premium" payments to attract labour etc. The APSAB index is based on an "Average PSA Building" compiled by analysing the bills of quantities for a range of representative contracts for typical PSA buildings into the work categories of the Formula Indices.

The percentage of the value of work in each of the work groups are as follows:

	%
A Demolitions, excavations, earthwork and piling	9.24
B Concrete work	22.20
C Brickwork and blockwork	9.25
D Masonry	0.24
E Roofing and cladding	4.12
F Carpentry and joinery	6.57
G Structural steelwork (unframed)	0.56
H Metalwork	4.73
I Plumbing	3.16
J Floor, wall and ceiling finishes	5.79

K Glazing	0.94
L Decorations	1.76
M External Works	2.71
Framed steelwork	3.75
Heating, hot water and ventilating installation	15.00
Electrical installation	10.00
	100.00

The APSAB index is calculated by applying the monthly Building Formula indices to the work categories of the 'Average PSA Building'. In effect, the index measures the movement in the cost of building the whole 'average building' each month. The specialist engineering indices are calculated from the monthly indices (E1 and E2) for Electrical Installations formula and indices (H1 and H2) for Heating etc. installations. These are weighted 40 per cent labour to 60 per cent materials for heating etc. and 57 per cent labour to 43 per cent materials for electrical.

The index is appropriate for any contract with an average balance of trades. For contracts with a disproportionate amount of work in any trade, a separate calculation should be made by application of the appropriate building formula indices.'

Commentary

The indices are broadly representative of the movement of building costs as measured by the formula and can be regarded as factor cost indices. The PSA append the following warning to their technical memorandum:

'It must be understood that the figures are related to costs as measured by increases in the costs of materials (Department of Trade and Industry indices) and labour (based on the National Working Rule Agreement) and have not borne any relation to tender prices charged by the industry.'

From that statement it is clear that they *do not* make any allowance for market conditions. Their main use, therefore, is in making an assessment of the change in cost to a contractor, for all or one particular factor, between one time period or another.

Publications

(a) Data Source

PSA *Quantity Surveyors Information Notes* prepared by the Directorate of Building and Quantity Surveying Services Division of the PSA Specialist Services, available on subscription from Publications Sales, Building Research Establishment, Garston, Watford, WD2 7JR.

(b) Description of Methodology

Details of the methodology are not freely available but are broadly as described above.

Table 6.8.1 Movement of Building Costs by Application of the PSA Price Adjustment Formula (APSAB Index) - Building

Base: 1985 = 100*

Year	Q1	Q2	Q3	Q4	Average
1970	16	16	16	17	16
1971	17	18	18	18	18
1972	18	18	19	21	19
1973	21	21	23	23	22
1974	24	26	28	29	27
1975	31	32	35	35	33
1976	37	38	41	42	40
1977	44	45	47	48	46
1978	48	49	51	52	50
1979	53	55	60	62	58
1980	64	66	72	73	69
1981	73	75	77	79	76
1982	80	82	84	85	83
1983	85	87	89	90	88
1984	91	92	95	96	94
1985	97	98	101	101	99
1986	101	101	103	104	102
1987	104	105	108	108	106
1988	109	111	114	116	113
1989	117	120	124	124**	121**
1990					
1991					
1992					
1993					
1994					
1995					

Source: PSA *Quantity Surveyors Information Notes.*

* Indices from 1970 to 1983 inclusive converted arithmetically by the authors from bases of 1970 = 100, 1975 = 100, and 1980 = 100 to 1985 = 100.

** Provisional.

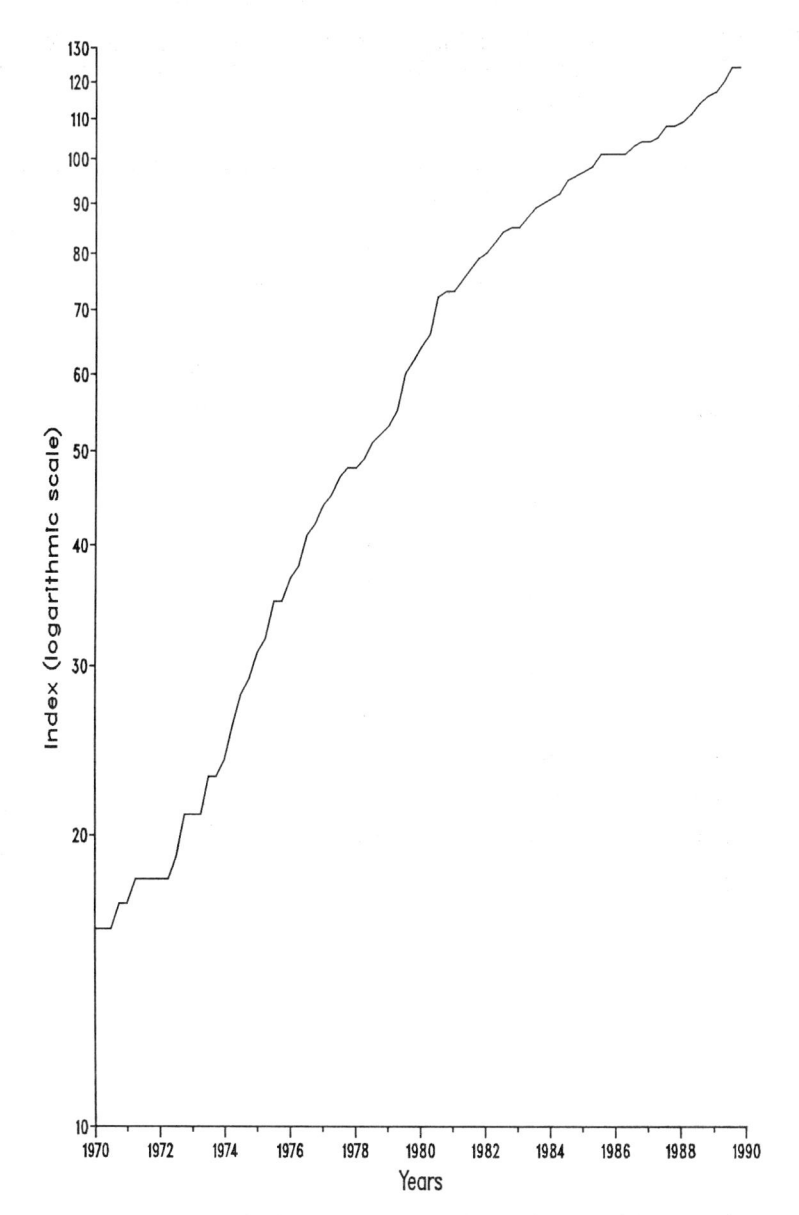

Figure 6.8.1 Movement of Building Costs Measured by Application of the PSA Price Adjustment Formula (APSAB Index) – Building (1985 = 100), 1970 Q1 – 1989 Q4

Table 6.8.2 Movement of Building Costs by Application of the PSA Price Adjustment Formula (APSAB Index) - H & V and Air Conditioning

Base: 1985 = 100*

Year	Q1	Q2	Q3	Q4	Average
1975	31	33	34	35	33
1976	35	37	38	40	38
1977	41	42	43	45	43
1978	46	47	48	51	48
1979	53	54	59	62	57
1980	67	69	70	71	69
1981	75	77	78	79	77
1982	82	84	84	85	84
1983	86	88	90	90	89
1984	91	93	94	94	93
1985	97	99	99	99	99
1986	100	103	103	104	103
1987	105	108	109	110	108
1988	111	114	115	117	114
1989	118	122	123	125**	122**
1990					
1991					
1992					
1993					
1994					
1995					

Source: PSA *Quantity Surveyors Information Notes*.

* Indices from 1970 to 1983 inclusive converted arithmetically by the authors from bases of 1975 = 100, and 1980 = 100 to 1985 = 100.

** Provisional.

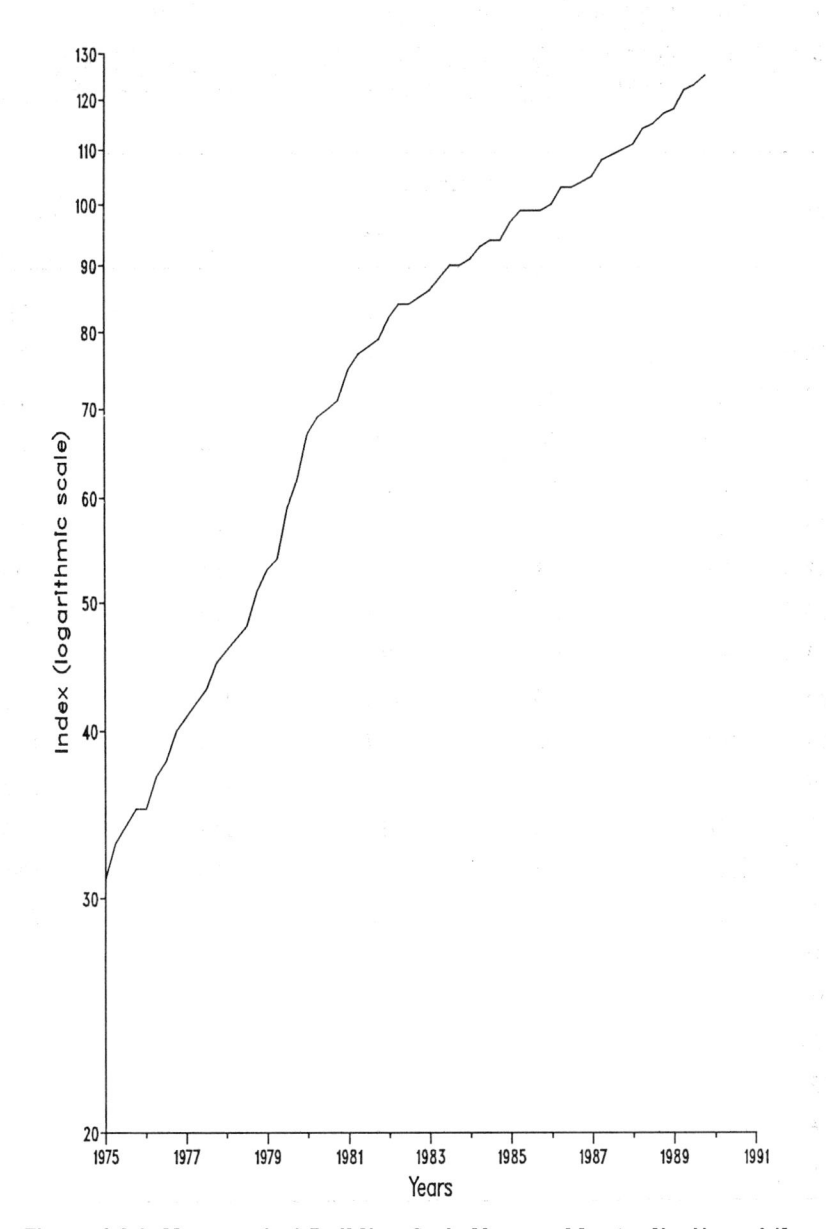

Figure 6.8.2 Movement of Building Costs Measured by Application of the PSA Price Adjustment Formula (APSAB Index) – H & V and Air Conditioning (1985 = 100), 1975 Q1 – 1989 Q4

Table 6.8.3 Movement of Building Costs by Application of the PSA Price Adjustment Formula (APSAB Index) - Electrical

Base: 1985 = 100*

Year	Q1	Q2	Q3	Q4	Average
1975	32	33	34	34	33
1976	36	38	39	40	38
1977	42	43	44	45	44
1978	49	50	50	52	50
1979	56	57	59	61	58
1980	66	69	72	77	71
1981	79	79	80	81	80
1982	82	89	90	90	88
1983	90	96	97	98	95
1984	99	100	102	102	101
1985	104	105	107	109	106
1986	110	110	111	112	111
1987	116	116	117	119	117
1988	124	124	124	126	125
1989	131	131	132	134**	132**
1990					
1991					
1992					
1993					
1994					
1995					

Source: PSA *Quantity Surveyors Information Notes.*

*   Indices from 1970 to 1983 inclusive converted arithmetically by the authors from bases of 1975 = 100, and 1980 = 100 to 1985 = 100.

**   Provisional.

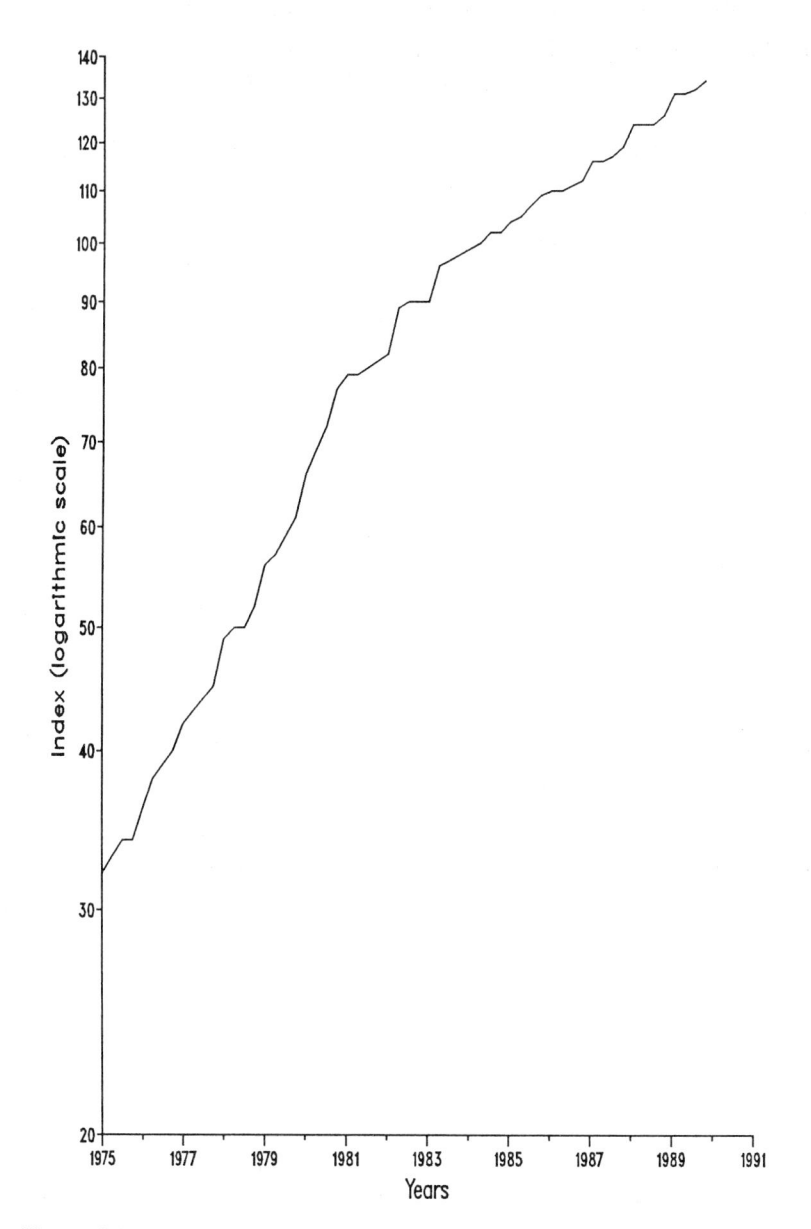

Figure 6.8.3 Movement of Building Costs Measured by Application of the PSA Price Adjustment Formula (APSAB Index) – Electrical (1985 = 100), 1975 Q1 – 1989 Q4

Table 6.8.4 Movement of Building Costs by Application of the PSA Price Adjustment Formula (APSAB Index) - Combined Index

Base: 1985 = 100*

Year	Q1	Q2	Q3	Q4	Average
1975	31	32	35	35	33
1976	36	38	41	42	39
1977	43	45	47	47	46
1978	48	49	51	52	50
1979	53	55	60	62	58
1980	64	66	72	73	69
1981	74	75	77	79	76
1982	80	82	84	85	83
1983	85	87	90	91	88
1984	92	93	96	96	94
1985	97	99	101	101	100
1986	102	102	104	105	103
1987	105	107	109	110	108
1988	111	113	115	117	114
1989	118	121	124	125**	122**
1990					
1991					
1992					
1993					
1994					
1995					

Source: PSA *Quantity Surveyors Information Notes.*

*   Indices from 1970 to 1983 inclusive converted arithmetically by the authors from bases of 1975 = 100, and 1980 = 100 to 1985 = 100.

**   Provisional.

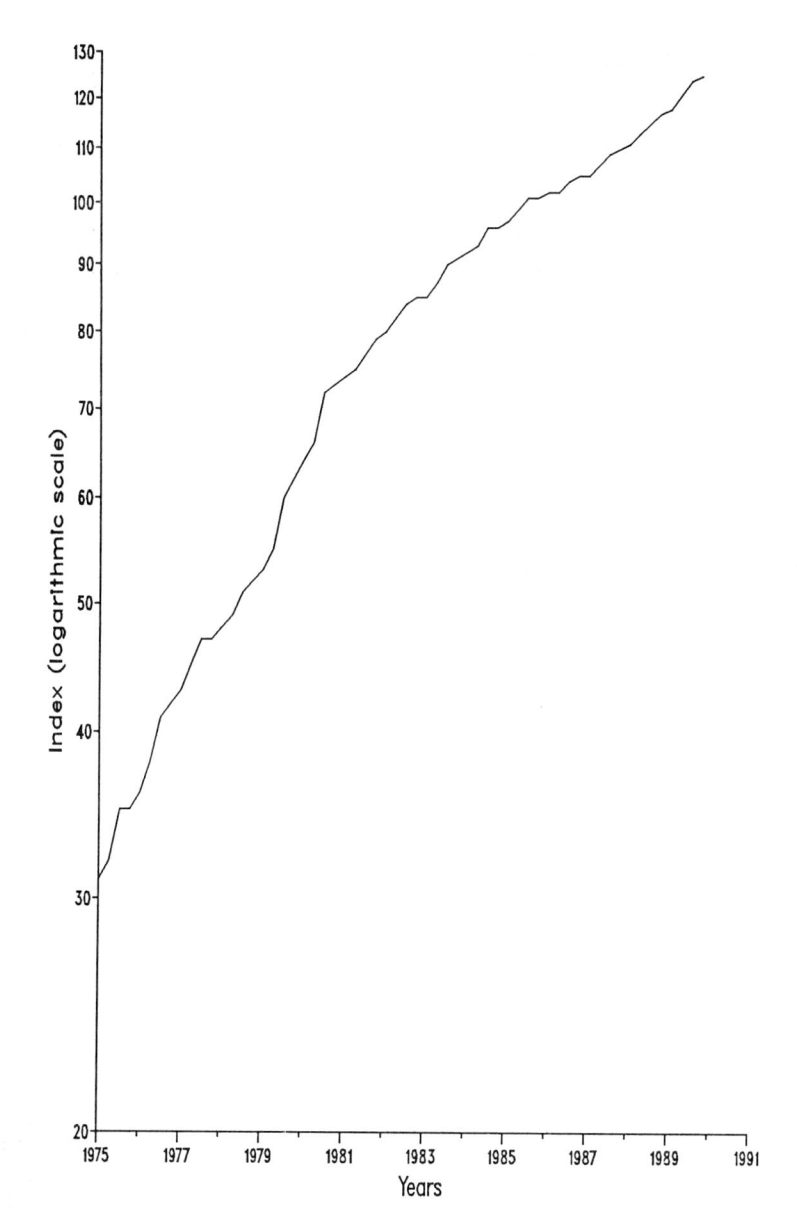

Figure 6.8.4 Movement of Building Costs Measured by Application of the PSA Price Adjustment Formula (APSAB Index) – Combined Index (1985 = 100), 1975 Q1 – 1989 Q4

7

Comparison and Review of Current Indices

Part B has so far dealt individually with indices in general use. These indices are now compared in detail and a review is made of the evidence they provide about the extent of cost and price increases in construction and how this evidence compares with inflation in the economy in general.

Turning to the comparison of the construction indices, they can be separated into four groups, as follows:

(a) Those which are compiled from information obtained from bills of quantities and which attempt to measure the change in the level of the cost to the client from one time period to another - tender price indices.
(b) Those which attempt to measure the change in the level of the cost to the client of construction *completed* in a given quarter from one time to another, and do not monitor the changes in the levels of tenders - output price indices.
(c) Those which are based on factor costs and attempt to measure the change in the level of the theoretical cost to the contractor from one time period to another - building cost indices.
(d) Those which are the same as (c) - building costs - but are concerned with subsections of whole buildings.

The following four tables and graphs compare the indices within the four groups and the fifth table and graph compare them all. All the indices for this purpose have been converted to 1985 = 100 as a base for this comparison. The BCIS tender price index used in the comparison is the all-in one.

The first group (Table 7.1, Figure 7.1) compares the following tender price indices:

1. DOE all-in public sector building TPI.
2. PSA QSSD all-in index of building tender prices.
3. DOE all-in road construction TPI.
4. BCIS all-in TPI.
5. D L & E TPI.
6. DOE price index of public sector housebuilding.

It will be seen that there is very little difference between the first, second and fourth indices over the period covered. They have troughs and peaks from 1975 onwards which roughly correlate, whereas the others show some differences. The DOE road construction tender price index seems to have a trough in 1976, but had caught up by the end of 1979, and was then ahead of them all, with the odd exception, from 1980 until 1984; the D L & E tender price index had a slower rate of increase from 1975, but, since 1986 has pulled ahead of all the rest; and the DOE price index of public sector housebuilding (PIPSH) which was at a lower level in the earlier years. These differences can be put down to differences in what is actually being measured. The DOE road construction tender price index (index 3) is unique in applying only to roads and the DOE price index of public sector housebuilding (index 6) applies *only* to housing, while index (1) and index (2) specifically *exclude* housing. Index (5) relates only to Greater London. In addition to differences of coverage by type of work or geographical area, noted above, there are differences according to the coverage of work in the public and private sectors: index (4) and index (5) cover both sectors but the other indices are restricted to public sector schemes alone. Further, the input of fluctuating and firm price tenders in each index also differs.

Another cause of divergence for some of the indices is the fact that where variants of an index are produced, for instance, alternative unweighted and value-weighted series (see section 5.1), a common (value-weighted) base is retained for *both* series. So, although the base date is stated as 1985 = 100, the actual figures for some series may not appear internally consistent with that base.

The second group of indices, (Table 7.2, Figure 7.2), attempts to measure the change in the level of the cost to the client of construction completed in a given quarter - output price indices. It will be seen that they have very small differences when compared with each other. Apart from both private and public housing, which lagged behind in the middle of the period, all these indices behaved in a similar way. It is notable, however, that the level of prices for work completed in private sector housing has increased at a faster rate than the rest since 1985, the base date.

The third group of indices, (Table 7.3, Figure 7.3), is based on factor costs and attempts to measure the change in the theoretical cost to the contractor - building cost indices. It is theoretical because no account is taken of attraction money paid to workers or materials bought at a premium in times of boom. It can be seen that the indices are almost identical. This is not surprising inasmuch as the labour and material inputs to construction are much the same for a wide range of work. The indices show that they are not very sensitive to the differences in composition and associated weights appropriate to the different categories of work defined here.

The fourth group of indices, (Table 7.4, Figure 7.4), is also based on factor costs but attempts to measure the change in theoretical cost to the contractor of subsections of whole buildings. These also show that over the period covered there was no great difference. The relative costs of electrical and mechanical services have see-sawed over the period covered but have increased at a very similar rate.

Consideration having been given to the indices in groups they can now be considered together (Table 7.5, Figure 7.5).

For clarity, only four indices, representative of their type, have been shown in Figure 7.5, one from each group of indices. It can be seen that there is a blank area between the tender price indices and the factor cost indices prior to 1983 with the tender price indices at a higher level (assuming that 1985 is correct for parity between the indices). This area does not represent total profit to the contractor. It represents the attraction money paid to workmen and prices paid at a premium for materials, in addition to a certain amount of increased profit during boom periods. Since 1987 the tender-based indices have gone ahead of factor-cost indices again. Provided, again, that 1985 is correct for parity, both types of indices should have progressed more or less in parallel until another boom period, when tender levels would probably once again outstrip factor cost levels. Although the latest figures suggest a boom period with tender price indices ahead of factor cost indices, it is felt that this is only a temporary situation. With regard to the output price indices, these have moved in a different manner from both factor cost indices and tender price indices due entirely to the differences in what is measured and the method of compilation.

It is emphasised that graph lines only relate to a particular index and reflect the changes in that index. If an index is consistently above another on a graph it does not mean that the work covered by the former is more costly than the latter. It simply means that the costs or prices measured by one index have changed *proportionately* more or less than those measured by the other index. In other words, it is important not to confuse differences in *absolute* cost or price levels with *relative* rates of change in those levels. All of the indices discussed and presented in this book measure the latter.

There have been occasions when the same project, having been included as a constituent tender in the calculation of two tender price indices, results in the individual 'project index' being above the average on one index and below the average on the other. For a project to differ from the average, possibly in opposite directions on the two indices, will often be due to the fact that the statistical population on which the indices are based are different. For example, should the scheme have been indexed in both the BCIS tender price index (section 5.3) and the QSSD index of building tender prices (section 5.2), it may be found to be higher than the average on one index and lower than the average on the other index. This would be because the BCIS tender price index population covers all building tenders whereas for the QSSD index of building tender prices the population covers mainly these tenders for new work accepted by the Department of the Environment. The two populations are different, one includes projects from both the public and private sectors whereas the other is entirely from the public sector. In addition, depending upon which index is being compared, the mix of firm price and variation of price tenders would also differ. It is differences such as these which would account for the apparent anomaly.

Another factor that would affect the foregoing comparisons is the choice of base date. Clearly all indices will be at the same apparent level on the chosen base date. This would mean that they would all converge at this date and could distort year-to-year comparisons if an 'unusual' year is taken as the base year for one or more of the indices.

Finally, we turn to summarise the evidence about construction cost and price trends and to compare it with price trends in the economy in general. The evidence given by the indices covering construction work in general (though some, it should be noted exclude housing) is as follows:

Index	1989 Index Number (1974 = 100)
Output prices	
DOE construction OPI - all new construction (Table 4.6)	381
Tender prices	
BCIS all-in TPI (Table 5.3.1)	337
Building costs	
BCIS general BCI (Table 6.1.1)	459
Spon's BCI (Table 6.2.1)	455

Over the period between 1974 and 1989, the output price and tender price indices suggest that prices rose by around 3.8 and 3.4 times the 1974 level respectively. The general building cost indices suggest that costs rose at a faster rate over the period and that by 1989 costs were around 4.6 times higher than in 1974.

Three indices for price trends in the economy as a whole, covering the same period, are given in Appendix A (Tables A2-A4) for retail prices, total home costs (implied GDP deflator) and capital goods prices. The evidence they give for the period 1974-1989 is as follows:

Index	1989 Index (1974 = 100)
Retail prices (Appendix Table A2)	419
Total home costs (Appendix Table A3)	419
Capital goods prices (Appendix Table A4)	386

These three series suggest that, by 1989, prices were around 4.2 times their 1974 level for retail prices and total home costs but somewhat less than four times higher for capital goods (3.9).

In summary, therefore, it would appear that building tender prices and output prices have risen somewhat less than prices in general but that building *costs* have risen rather more.

Comparison and review of current indices

Table 7.1 Comparison of indices based on information obtained from bills of quantities

Base: 1985 = 100

Year	Qtr	DOE All-in public sector building tender price index	PSA QSSD All-in index of building tender prices	DOE All-in road construction tender price index	BCIS All-in tender price index	DL&E tender price index	DOE price index of public sector house building All-in
1975	(1)	43	45	44	43	46	42
	(2)	45	46	44	42	47	42
	(3)	44	46	43	43	45	43
	(4)	43	45	41	43	46	45
1976	(1)	45	47	41	46	44	44
	(2)	46	48	40	45	45	46
	(3)	48	50	40	47	47	46
	(4)	47	49	41	48	47	46
1977	(1)	49	51	45	50	48	50
	(2)	52	54	47	54	48	50
	(3)	54	54	50	55	50	52
	(4)	56	58	53	53	50	53
1978	(1)	57	57	57	56	52	55
	(2)	61	60	60	60	53	57
	(3)	64	64	61	63	58	58
	(4)	63	68	64	67	64	62
1979	(1)	68	72	67	71	65	62
	(2)	76	76	72	74	67	66
	(3)	81	83	76	82	73	71
	(4)	83	91	88	87	77	74
1980	(1)	86	91	96	88	82	81
	(2)	94	101	100	92	92	84
	(3)	89	93	101	94	88	87
	(4)	86	97	96	89	86	88
1981	(1)	83	85	94	87	91	87
	(2)	87	89	91	87	88	84
	(3)	85	86	89	86	87	84
	(4)	86	90	87	84	89	85

Table 7.1 continued

Year	Qtr	DOE All-in public sector building tender price index	PSA QSSD All-in index of building tender prices	DOE All-in road construction tender price index	BCIS All-in tender price index	DL&E tender price index	DOE price index of public sector house building All-in
1982	(1)	88	89	88	89	88	88
	(2)	84	88	91	88	86	87
	(3)	86	85	95	86	89	88
	(4)	87	84	96	87	89	89
1983	(1)	87	89	97	88	91	91
	(2)	87	86	95	88	92	94
	(3)	90	88	94	88	91	96
	(4)	91	89	95	90	92	94
1984	(1)	94	90	95	94	94	96
	(2)	95	94	96	96	94	96
	(3)	93	93	95	95	98	96
	(4)	93	88	99	96	99	97
1985	(1)	95	93	98	97	99	97
	(2)	98	99	100	102	100	99
	(3)	99	98	97	99	100	101
	(4)	99	98	102	103	101	104
1986	(1)	101	99	97	101	101	105
	(2)	101	101	93	102	104	103
	(3)	98	97	94	103	107	104
	(4)	104	100	95	105	107	103
1987	(1)	106	102	101	108	111	109
	(2)	110	106	106	106	114	112
	(3)	109	106	107	109	121	114
	(4)	115	109	111	117	128	117
1988	(1)	121	108	110	119	132	120
	(2)	125	122	114	123	137	124
	(3)	129	127	117	128	147	125
	(4)	127	128	116	128	150	132
1989	(1)	134	133	130	135	156	134
	(2)	135	143†	125**	135	153	138
	(3)	136	135	129**	141**	156	136
	(4)	130**			142**	158	137**

Table 7.1 continued

Year	Qtr	DOE All-in public sector building tender price index	PSA QSSD All-in index of building tender prices	DOE All-in road construction tender price index	BCIS All-in tender price index	DL&E tender price index	DOE price index of public sector house building All-in
1990	(1)					151**	
	(2)						
	(3)						
	(4)						

Source: Various, as previous tables

** Provisional.
† Based on a very small sample.

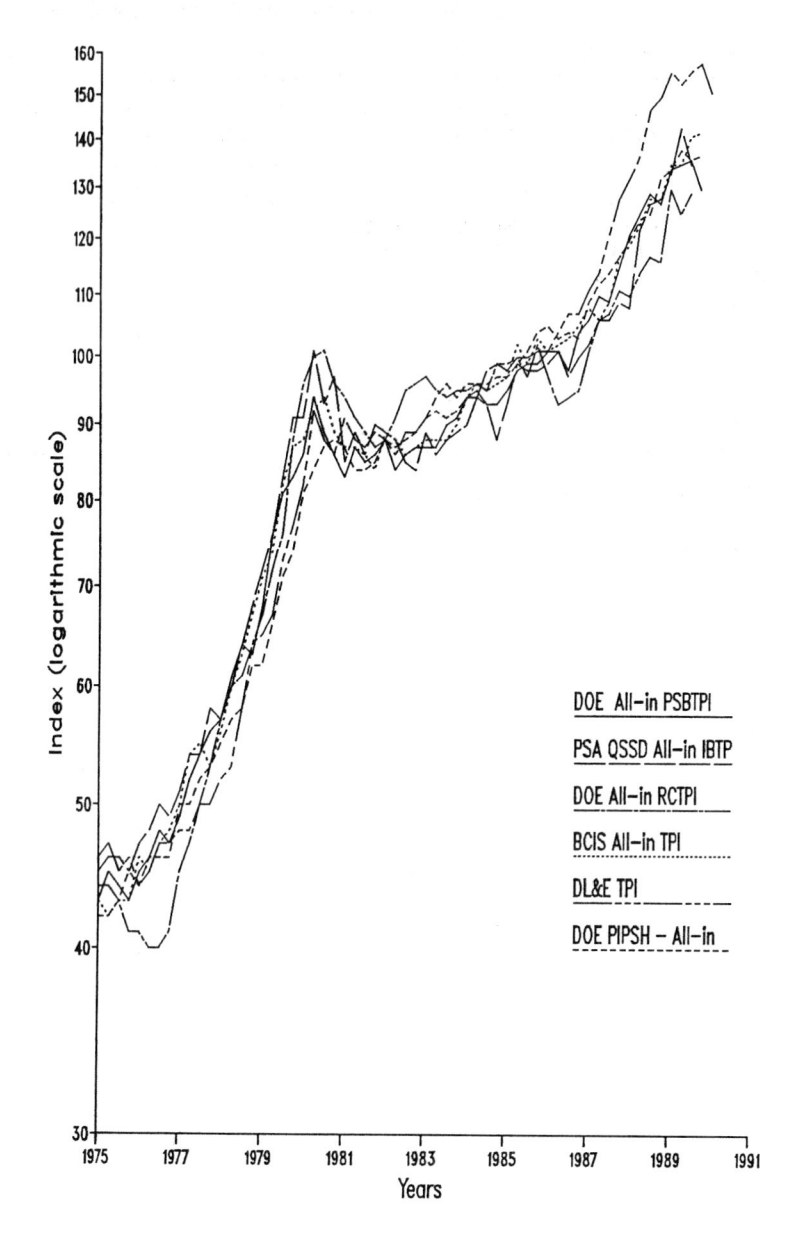

Figure 7.1 Comparison of Indices Based on Tender Prices (1985 = 100) 1975 Q1 – 1989 Q4

Table 7.2 Comparison of indices which attempt to measure the change in the level of the cost to the client of construction completed in a given quarter and do not monitor the changes in the levels of tenders

Base: 1985 = 100

DOE Construction Output Price Indices

Date Year Qtr	Public housing	Private housing	Public works	Private industrial	Private commercial	All new construction
1975 (1)	43	36	42	43	45	42
(2)	44	37	44	44	46	43
(3)	47	38	47	45	48	45
(4)	47	40	48	46	48	46
1976 (1)	48	40	49	46	49	46
(2)	48	40	50	48	49	47
(3)	50	42	51	49	51	49
(4)	51	42	51	50	52	50
1977 (1)	52	43	51	51	53	50
(2)	52	44	51	51	54	51
(3)	53	46	53	53	55	52
(4)	53	47	54	55	57	53
1978 (1)	54	48	55	56	57	54
(2)	56	50	57	57	59	56
(3)	58	53	60	60	62	59
(4)	60	55	62	63	65	61
1979 (1)	62	57	65	66	69	64
(2)	64	60	69	70	73	67
(3)	70	65	74	78	79	74
(4)	73	68	78	83	85	77
1980 (1)	77	72	84	90	90	82
(2)	80	76	89	93	93	86
(3)	87	80	97	99	99	92
(4)	89	82	99	100	100	94
1981 (1)	90	83	98	99	100	94
(2)	91	84	97	98	99	94
(3)	91	83	97	97	97	93
(4)	91	82	97	95	96	92
1982 (1)	90	82	96	94	95	91
(2)	90	83	96	95	95	91
(3)	90	84	96	95	96	92
(4)	90	84	95	94	95	92

Table 7.2 continued

Date Year Qtr	Public housing	Private housing	Public works	Private industrial	Private commercial	All new construction
1983 (1)	90	86	94	93	94	92
(2)	92	88	95	93	94	92
(3)	94	91	96	94	94	93
(4)	95	92	95	93	93	93
1984 (1)	96	93	95	95	93	94
(2)	97	94	95	97	93	95
(3)	98	96	97	97	94	96
(4)	98	96	97	94	95	96
1985 (1)	99	97	98	95	96	97
(2)	99	99	99	100	98	99
(3)	101	101	101	103	102	101
(4)	101	103	102	102	104	103
1986 (1)	102	105	103	102	104	104
(2)	103	108	102	102	105	104
(3)	104	111	102	100	107	105
(4)	104	113	102	99	108	106
1987 (1)	105	116	104	101	109	108
(2)	106	119	106	105	109	110
(3)	108	123	108	108	110	113
(4)	110	128	110	109	113	115
1988 (1)	113	133	113	111	116	119
(2)	116	139	116	113	120	122
(3)	119	146	120	116	124	127
(4)	122	152	124	120	128	132
1989 (1)	125	158	129	124	131	136
(2)	129	163	132	127	135	140
(3)	132**	167**	135**	132**	137**	143**
(4)	134**	170**	137**	135**	138**	145**
1990 (1)						
(2)						
(3)						
(4)						

Source: Various, as previous tables.

** Provisional.

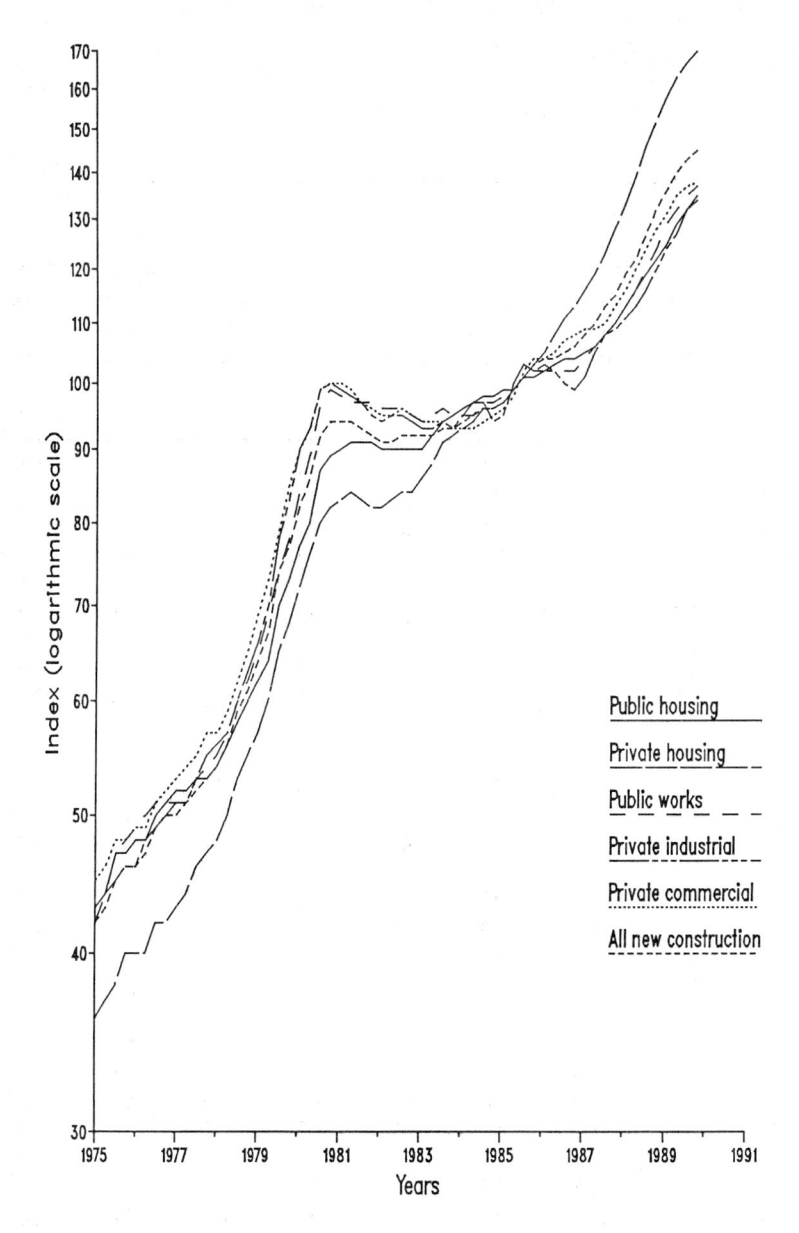

Figure 7.2 Comparison of Indices Measuring Output Prices (1985 = 100) 1975 Q1 – 1989 Q4

Table 7.3 Comparison of indices based on factor costs which attempt to measure the change in the theoretical cost to the contractor

Base: 1985 = 100

Date Year Qtr	General building cost (exc. M&E) index	General building cost index	Steel framed construction cost index	Concrete framed construction cost index	Brick construction cost index	Spon's building cost index	Building housing cost index	PSA price adj. form. appl. APSAB
1975(1)	31	31	31	31	31	31	33	31
(2)	33	33	33	33	33	33	34	32
(3)	35	35	35	35	35	36	34	32
(4)	36	36	36	36	35	36	38	35
1976(1)	37	37	37	36	37	37	38	37
(2)	38	38	38	38	38	39	40	38
(3)	42	41	41	41	41	42	42	41
(4)	43	42	42	42	42	43	43	42
1977(1)	44	44	44	43	44	44	44	44
(2)	46	45	45	45	45	45	46	45
(3)	48	46	47	46	47	47	47	47
(4)	48	47	48	47	47	47	48	48
1978(1)	49	49	49	48	48	47	48	48
(2)	50	49	50	49	49	48	49	49
(3)	52	51	52	51	51	51	51	51
(4)	53	53	53	52	52	52	53	52
1979(1)	54	54	54	54	54	53	54	53
(2)	56	55	56	55	55	54	55	55
(3)	61	60	60	60	60	60	61	60
(4)	63	62	63	63	62	62	63	62
1980(1)	64	65	65	65	65	63	65	64
(2)	67	67	67	67	67	65	67	66
(3)	73	72	72	72	72	72	73	72
(4)	74	73	73	73	73	73	74	73
1981(1)	74	74	74	74	74	73	75	73
(2)	76	76	76	76	76	74	77	75
(3)	78	78	77	78	78	78	79	77
(4)	80	79	79	79	79	80	81	79

Table 7.3 continued

BCIS Building Cost Indices

Date Year Qtr	General building cost (exc. M&E) index	General building cost index	Steel framed construction cost index	Concrete framed construction cost index	Brick construction cost index	Spon's building cost index	Building housing cost index	PSA price adj. form. appl. APSAB
1982(1)	81	81	81	81	81	82	82	80
(2)	83	83	83	83	83	83	84	82
(3)	85	85	85	85	85	86	86	84
(4)	86	85	85	86	86	87	87	85
1983(1)	86	86	85	86	86	87	87	85
(2)	88	88	88	88	88	88	89	87
(3)	91	90	90	91	91	91	92	89
(4)	91	91	91	91	92	92	92	90
1984(1)	92	92	92	92	92	93	93	91
(2)	93	94	93	94	94	93	94	92
(3)	96	96	96	96	96	96	96	95
(4)	97	96	96	96	97	97	96	96
1985(1)	98	98	98	98	98	98	97	97
(2)	99	99	99	99	99	99	99	98
(3)	102	101	101	101	101	101	102	101
(4)	102	102	102	102	102	102	102	101
1986(1)	102	102	102	102	102	103	102	101
(2)	103	103	103	103	103	104	103	101
(3)	105	105	105	105	105	107	106	103
(4)	106	106	106	105	106	107	107	104
1987(1)	106	107	107	106	107	109	108	104
(2)	108	108	108	107	108	109	109	105
(3)	110	110	111	110	111	113	112	108
(4)	111	111	111	110	111	113	113	108
1988(1)	111	112	112	111	112	114	114	109
(2)	113	114	114	113	114	115	116	111
(3)	117	117	117	116	117	120	119	114
(4)	118	118	118	118	118	122	121	116

Table 7.3 continued

BCIS Building Cost Indices

Date Year Qtr	General building cost (exc. M&E) index	General building cost index	Steel framed construction cost index	Concrete framed construction cost index	Brick construction cost index	Spon's building cost index	Building housing cost index	PSA price adj. form. appl. APSAB
1989(1)	119	120	120	120	120	123	122	117
(2)	122	122	123	122	122	124	124	120
(3)	126	126	126	126	126	130	130	124
(4)	127	127	127	127	127	130	131	124**
1990(1)	128**	128**	128**	128**	128**	131**	132**	
(2)								
(3)								
(4)								
1991(1)								
(2)								
(3)								
(4)								
1992(1)								
(2)								
(3)								
(4)								
1993(1)								
(2)								
(3)								
(4)								
1994(1)								
(2)								
(3)								
(4)								
1995(1)								
(2)								
(3)								
(4)								

Source: Various, as previous tables.

** Provisional.

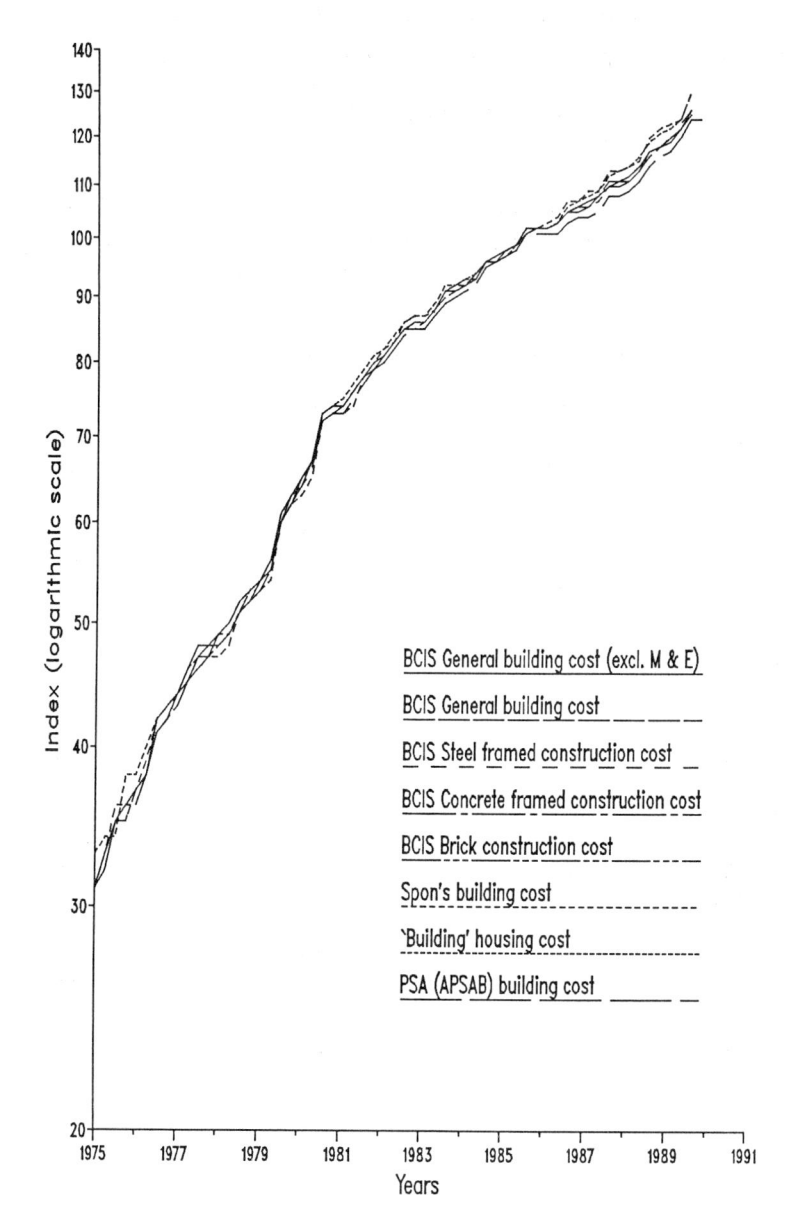

Figure 7.3 Comparison of Indices Based on Factor Costs (1985 = 100) 1975 Q1 – 1989 Q4

Table 7.4 Comparison of indices based on factor costs which attempt to measure the change in the theoretical cost to the contractor of subsections of whole buildings

Base: 1985 = 100

Date Year	Qtr	BCIS mechanical and electrical engineering cost index	Spon's mechanical services cost index	Spon's electrical services cost index
1975	(1)	31	34	33
	(2)	33	35	33
	(3)	33	37	35
	(4)	33	37	35
1976	(1)	35	39	39
	(2)	37	41	41
	(3)	38	44	42
	(4)	40	46	42
1977	(1)	41	47	45
	(2)	42	48	45
	(3)	43	48	45
	(4)	44	49	45
1978	(1)	47	51	50
	(2)	48	52	52
	(3)	49	54	53
	(4)	51	57	54
1979	(1)	53	59	59
	(2)	55	60	59
	(3)	58	64	60
	(4)	61	66	63
1980	(1)	66	70	69
	(2)	68	73	69
	(3)	70	73	72
	(4)	72	74	79
1981	(1)	75	77	80
	(2)	76	79	81
	(3)	77	80	81
	(4)	78	81	82
1982	(1)	81	84	83
	(2)	84	86	86
	(3)	84	86	86
	(4)	85	86	87

Table 7.4 continued

Date		BCIS mechanical and electrical engineering cost index	Spon's mechanical services cost index	Spon's electrical services cost index
Year	Qtr			
1983	(1)	86	87	88
	(2)	89	87	92
	(3)	91	89	92
	(4)	92	90	93
1984	(1)	93	91	93
	(2)	94	94	94
	(3)	96	95	97
	(4)	96	96	97
1985	(1)	98	99	98
	(2)	100	100	98
	(3)	101	100	101
	(4)	102	100	102
1986	(1)	103	101	103
	(2)	104	102	102
	(3)	105	102	102
	(4)	106	103	104
1987	(1)	108	104	110
	(2)	110	106	110
	(3)	111	107	111
	(4)	112	108	112
1988	(1)	114	109	117
	(2)	116	113	118
	(3)	117	114	119
	(4)	119	116	120
1989	(1)	121	117	125
	(2)	124	121	125
	(3)	126**	121	125
	(4)		122**	127**
1990	(1)			
	(2)			
	(3)			
	(4)			

Source: *BCIS Manual* and *Spon's Mechanical and Electrical Price Book.*

** Provisional

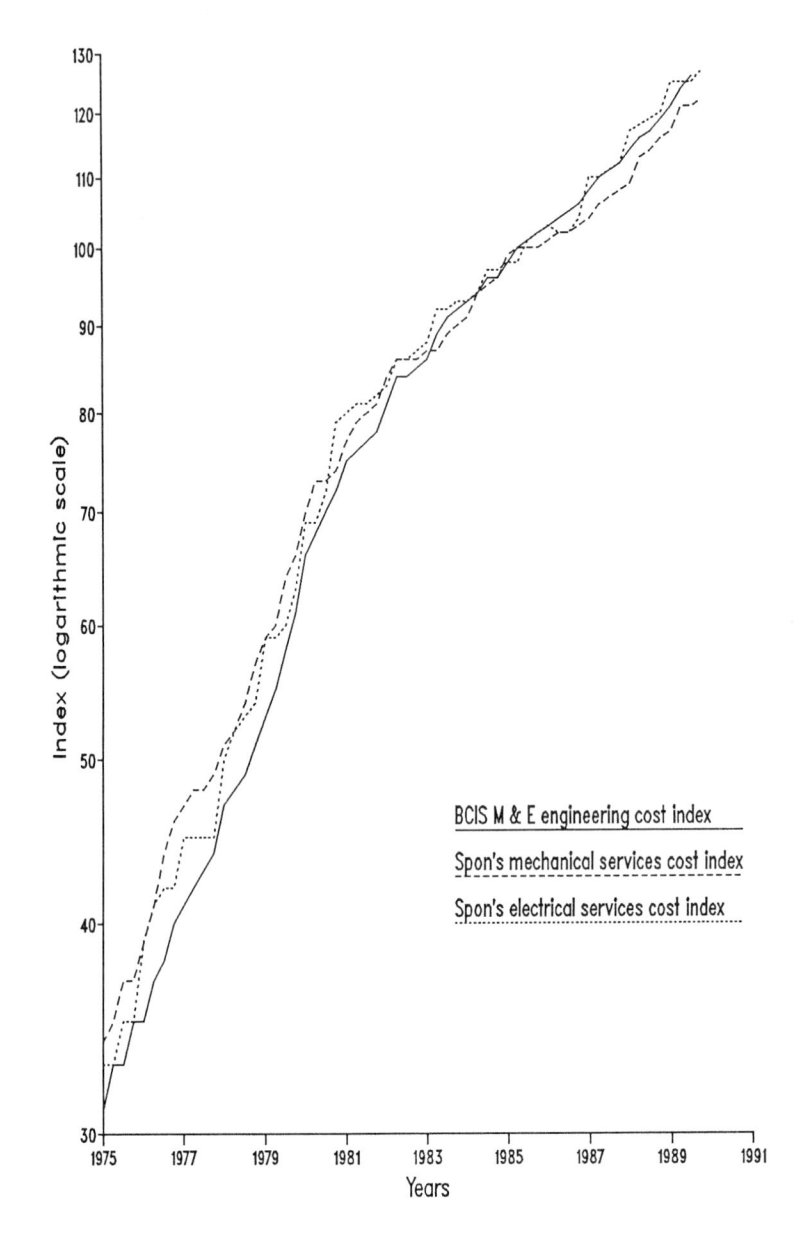

Figure 7.4 Comparison of Indices Based on Factor Costs for Sub-sections of Buildings (1985 = 100), 1975 Q1 – 1989 Q4

For Table 7.5, see overleaf.

Table 7.5 Direct comparison of the groups of indices shown in the four previous tables

Base: 1985 = 100

Year	Quarter	DOE A-i PSBTPI	PSA QSSD A-i IBTP	DOE A-i road const. TPI	BCIS A-in tender price index	DL & E tender price index	DOE PIPSHB A-i	Public housing	Private housing	Public works	Private industrial	Private commercial	All new construction
1975	(1)	43	45	44	43	46	42	43	36	42	43	45	42
	(2)	45	46	44	42	47	42	44	37	44	44	46	43
	(3)	44	46	43	43	45	43	47	38	47	45	48	45
	(4)	43	45	41	43	46	45	47	40	48	46	48	46
1976	(1)	45	47	41	46	44	44	48	40	49	46	49	46
	(2)	46	48	40	45	45	46	48	40	50	48	49	47
	(3)	48	50	40	47	47	46	50	42	51	49	51	49
	(4)	47	49	41	48	47	46	51	42	51	50	52	50
1977	(1)	49	51	45	50	48	50	52	43	51	51	53	50
	(2)	52	54	47	54	48	50	52	44	51	51	54	51
	(3)	54	54	50	55	50	52	53	46	53	53	55	52
	(4)	56	58	53	53	50	53	53	47	54	55	57	53
1978	(1)	57	57	57	56	52	55	54	48	55	56	57	54
	(2)	61	60	60	60	53	57	56	50	57	57	59	56
	(3)	64	64	61	63	58	58	58	53	60	60	62	59
	(4)	63	68	64	67	64	62	60	55	62	63	65	61
1979	(1)	68	72	67	71	65	62	62	57	65	66	69	64
	(2)	76	76	72	74	67	66	64	60	69	70	73	67
	(3)	81	83	76	82	73	71	70	65	74	78	79	74
	(4)	83	91	88	87	77	74	73	68	78	83	85	77
1980	(1)	86	91	96	88	82	81	77	72	84	90	90	82
	(2)	94	101	100	92	92	84	80	76	89	93	93	86
	(3)	89	93	101	94	88	87	87	80	97	99	99	92
	(4)	86	97	96	89	86	88	89	82	99	100	100	94
1981	(1)	83	85	94	87	91	87	90	83	98	99	100	94
	(2)	87	89	91	87	88	84	91	84	97	98	99	94
	(3)	85	86	89	86	87	84	91	83	97	97	97	93
	(4)	86	90	87	84	89	85	91	82	97	95	96	92
1982	(1)	88	89	88	89	88	88	90	82	96	94	95	91
	(2)	84	88	91	88	86	87	90	83	96	95	95	91
	(3)	86	85	95	86	89	88	90	84	96	95	96	92
	(4)	87	84	96	87	89	89	90	84	95	94	95	92

Base: 1985 = 100

GBC (excl. M&E)	GBCI	SFCCI	CFCCI	BCCI	Spon's building cost index	'Building' HCI	NEDO PAFA APSAB	BCIS M&E cost index	Spon's MSCI	Spon's ESCI	Year	Quarter
31	31	31	31	31	31	33	31	31	34	33	1975	(1)
33	33	33	33	33	33	34	32	33	35	33		(2)
35	35	35	35	35	36	34	32	33	37	35		(3)
36	36	36	36	35	36	38	35	33	37	35		(4)
37	37	37	36	37	37	38	37	35	39	39	1976	(1)
38	38	38	38	38	39	40	38	37	41	41		(2)
42	41	41	41	41	42	42	41	38	44	42		(3)
43	42	42	42	42	43	43	42	40	46	42		(4)
44	44	44	43	44	44	44	44	41	47	45	1977	(1)
46	45	45	45	45	45	46	45	42	48	45		(2)
48	46	47	46	47	47	47	47	43	48	45		(3)
48	47	48	47	47	47	48	48	44	49	45		(4)
49	49	49	48	48	47	48	48	47	51	50	1978	(1)
50	49	50	49	49	48	49	49	48	52	52		(2)
52	51	52	51	51	51	51	51	49	54	53		(3)
53	53	53	52	52	52	53	52	51	57	54		(4)
54	54	54	54	54	53	54	53	53	59	59	1979	(1)
56	55	56	55	55	54	55	55	55	60	59		(2)
61	60	60	60	60	60	61	60	58	64	60		(3)
63	62	63	63	62	62	63	62	61	66	63		(4)
64	65	65	65	65	63	65	64	66	70	69	1980	(1)
67	67	67	67	67	65	67	66	68	73	69		(2)
73	72	72	72	72	72	73	72	70	73	72		(3)
74	73	73	73	73	73	74	73	72	74	79		(4)
74	74	74	74	74	73	75	73	75	77	80	1981	(1)
76	76	76	76	76	74	77	75	76	79	81		(2)
78	78	77	78	78	78	79	77	77	80	81		(3)
80	79	79	79	79	80	81	79	78	81	82		(4)
81	81	81	81	81	82	82	80	81	84	83	1982	(1)
83	83	83	83	83	83	84	82	84	86	86		(2)
85	85	85	85	85	86	86	84	84	86	86		(3)
86	85	85	86	86	87	87	85	85	86	87		(4)

Table 7.5 continued

Year	Quarter	DOE A-i PSBTPI	PSA QSSD A-i IBTP	DOE A-i road const. TPI	BCIS A-in tender price index	DL & E tender price index	DOE PIPSHB A-i	Public housing	Private housing	Public works	Private industrial	Private commercial	All new construction
1983	(1)	87	89	97	88	91	91	90	86	94	93	94	92
	(2)	87	86	95	88	92	94	92	88	95	93	94	92
	(3)	90	88	94	88	91	96	94	91	96	94	94	93
	(4)	91	89	95	90	92	94	95	92	95	93	93	93
1984	(1)	94	90	95	94	94	96	96	93	95	95	93	94
	(2)	95	94	96	96	94	96	97	94	95	97	93	95
	(3)	93	93	95	95	98	96	98	96	97	97	94	96
	(4)	93	88	99	96	99	97	98	96	97	94	95	96
1985	(1)	95	93	98	97	99	97	99	97	98	95	96	97
	(2)	98	99	100	102	100	99	99	99	99	100	98	99
	(3)	99	98	97	99	100	101	101	101	101	103	102	101
	(4)	99	98	102	103	101	104	101	103	102	102	104	103
1986	(1)	101	99	97	101	101	105	102	105	103	102	104	104
	(2)	101	101	93	102	104	103	103	108	102	102	105	104
	(3)	98	97	94	103	107	104	104	111	102	100	107	105
	(4)	104	100	95	105	107	103	104	113	102	99	108	106
1987	(1)	106	102	101	108	111	109	105	116	104	101	109	108
	(2)	110	106	106	106	114	112	106	119	106	105	109	110
	(3)	109	106	107	109	121	114	108	123	108	108	110	113
	(4)	115	109	111	117	128	117	110	128	110	109	113	115
1988	(1)	121	108	110	119	132	120	113	133	113	111	116	119
	(2)	125	122	114	123	137	124	116	139	116	113	120	122
	(3)	129	127	117	128	147	125	119	146	120	116	124	127
	(4)	127	128	116	128	150	132	122	152	124	120	128	132
1989	(1)	134	133	130	135	156	134	125	158	129	124	131	136
	(2)	135	143†	125*	135	153	138	129	163	132	127	135	140
	(3)	136	135	129*	141*	156	136	132*	167*	135*	132*	137*	143*
	(4)	130*			142*	158	137*	134*	170*	137*	135*	138*	145*
1990	(1)					151*							
	(2)												
	(3)												
	(4)												

Source: Various, as previous tables.

† Based on very small samples. * Provisional.

GBC (excl. M&E)	GBCI	SFCCI	CFCCI	BCCI	Spon's building cost index	'Building' HCI	NEDO PAFA APSAB	BCIS M&E cost index	Spon's MSCI	Spon's ESCI	Year	Quarter
86	86	85	86	86	87	87	85	86	87	88	1983	(1)
88	88	88	88	88	88	89	87	89	87	92		(2)
91	90	90	91	91	91	92	89	91	89	92		(3)
91	91	91	91	92	92	92	90	92	90	93		(4)
92	92	92	92	92	93	93	91	93	91	93	1984	(1)
93	94	93	94	94	93	94	92	94	94	94		(2)
96	96	96	96	96	96	96	95	96	95	97		(3)
97	96	96	96	97	97	96	96	96	96	97		(4)
98	98	98	98	98	98	97	97	98	99	98	1985	(1)
99	99	99	99	99	99	99	98	100	100	98		(2)
102	101	101	101	101	101	102	101	101	100	101		(3)
102	102	102	102	102	102	102	101	102	100	102		(4)
102	102	102	102	102	103	102	101	103	101	103	1986	(1)
103	103	103	103	103	104	103	101	104	102	102		(2)
105	105	105	105	105	107	106	103	105	102	102		(3)
106	106	106	105	106	107	107	104	106	103	104		(4)
106	107	107	106	107	109	108	104	108	104	110	1987	(1)
108	108	108	107	108	109	109	105	110	106	110		(2)
110	110	111	110	111	113	112	108	111	107	111		(3)
111	111	111	110	111	113	113	108	112	108	112		(4)
111	112	112	111	112	114	114	109	114	109	117	1988	(1)
113	114	114	113	114	115	116	111	116	113	118		(2)
117	117	117	116	117	120	119	114	117	114	119		(3)
118	118	118	118	118	122	121	116	119	116	120		(4)
119	120	120	120	120	123	122	117	121	117	125	1989	(1)
122	122	123	122	122	124	124	120	124	121	125		(2)
126*	126*	126*	126*	125*	130	130	124	126*	121	125		(3)
					130	131	124*		122*	127*		(4)
					131*	132*					1990	(1)
												(2)
												(3)
												(4)

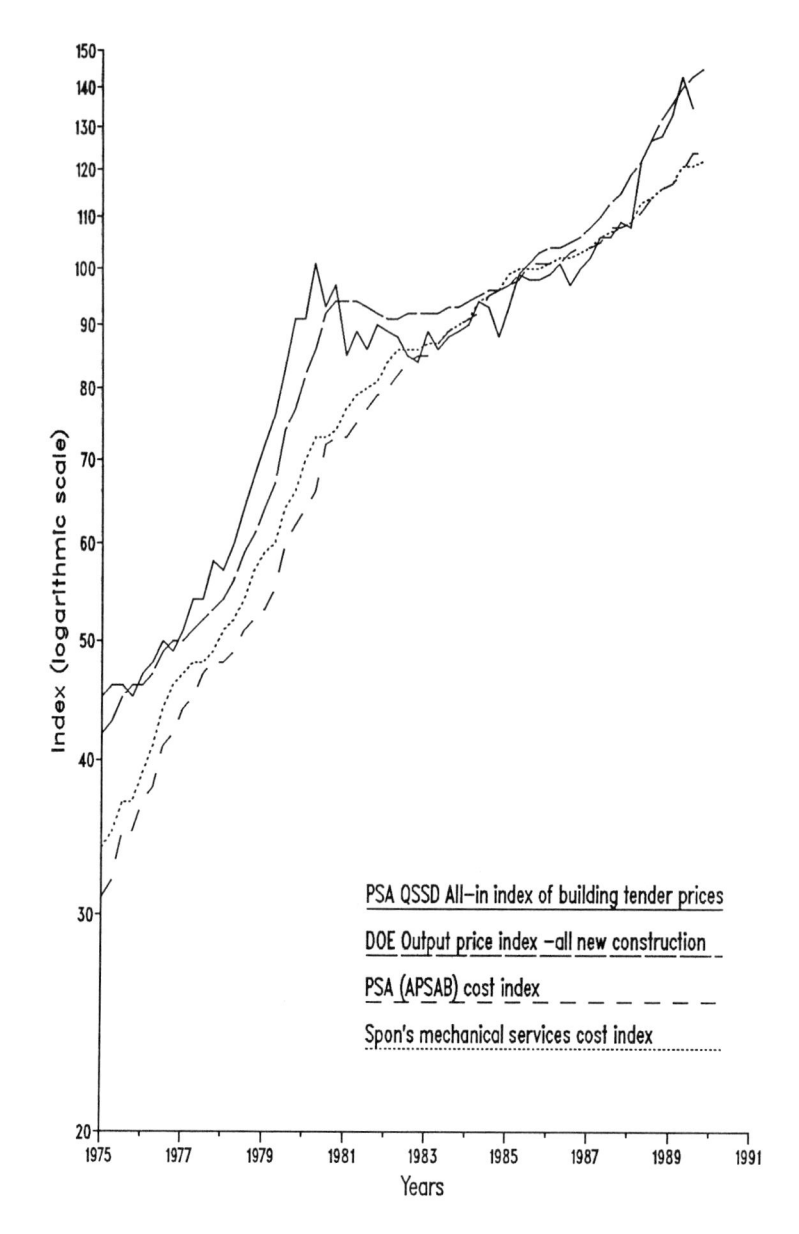

Figure 7.5 Comparison of Selected Indices from Tables 7.1 – 7.4 (1985 = 100) 1975 Q1 – 1989 Q4

Part C

Historical Construction Indices

8

Introduction to Historical Construction Indices

This part of the book brings together the main long-run historical series. As we have explained earlier, there is a particular need for indices covering long periods of time in construction because building and other construction works have long lengths of life, often - or even generally - exceeding 100 years (and in some cases, of course, very much longer still). The indices available have been devised by different people and organisations at different times for different purposes. These purposes are explained in the 'Commentary' on each index included in the next chapter but a brief summary statement is included at the end of this chapter. The series available also cover different periods of time. Ten series are included here, spanning (between them) the period from 1839/1850 to 1980.

An important problem facing the compiler of long-run historical series is that the further back in time one attempts to go, the less adequate the primary data available for constructing an index become. Some of the indices have to be compiled retrospectively using such information as happens to be available. Inevitably, this has disadvantages over currently-compiled series which may be based on specially designed systems for the collection and analysis of data and the contemporaneous calculation of index numbers. It will be appreciated, therefore, that many of the long-run historical series are less reliable than most of the currently-compiled indices. Another problem confronting the preparation of reliable long-run series, in particular, is the difficulty of allowing for the impact of technical change on prices and the influence of changing design, specification, standards and forms of construction over time. Consequently, the evidence provided by the long-run indices should be regarded as providing a broad indication of trends only; too much reliance should not be placed on year-to-year movements.

The user of long-run series, therefore, has to contend with additional problems over and above those confronting the user of current series. One solution to the problem of choice and the desire to reduce the magnitude of potential measurement errors, is to combine the evidence provided by several different indices by splicing them together. This is in fact, the solution adopted by the Central Statistical Office (CSO) for use in official statistics (see Sections 9.4 and 9.5 in Chapter 9). A long-run index compiled in this way, covering the period 1896-1974, was also a feature of the centenary edition of *Spon's Architects' and Builders' Price Book* (Davis, Belfield & Everest, 1975).

The description of the available long-run series, given in the following chapter, follows the same format as that used for the current series in chapters 5-7, but, in contrast, we do not group the long-run series according to type of index (output price index, tender price index, or cost index) as before, because the nature of the historical series available makes it inappropriate. This is because some of the indices, as described above, are based on a composite of both price and cost information, while other series (although purporting to measure price movements) are, in effect, cost indices, while others still are based on unspecified data and methods of calculation.

To conclude this introduction to the main historical series, we give a brief summary of the types of index available and the purposes for which they were prepared. A summary of the evidence they provide about long-term cost and price changes for construction, and a comparison with general price trends in the economy as a whole, is given at the end of this part of the book in chapter 10. The historical series covered in the next chapter are:

1. G.T.Jones's index of building prices, 1845-1922.
2. J.Saville's index of building prices, 1923-1939.
3. K.Maiwald's indices of costs for building and other construction, 1845-1938.
4. P.Redfern's indices of costs for building and works and for housing, 1839/1850-1953.
5. CSO CCA index of costs for new building works, 1888-1956.
6. Board of Inland Revenue (BIR) index of the construction costs of industrial buildings, 1896-1956.
7. Venning Hope cost of building index, 1914-1975.
8. Banister Fletcher's index of the comparative cost of building, 1920-1939.
9. BRS measured work index, London area, 1939-1969 Q2.
10. DOE cost of new construction (CNC) index, 1949 Q1 - 1980 Q1

With regard to the three broad types of index distinguished earlier (output price indices, tender price indices and cost indices), the DOE CNC index (index 10) was devised as an output price index for use in revaluing statistics of output and expenditure on construction work, but on the basis of factor cost information. The Venning Hope index (index 6) is described as a 'cost of building index' but is in fact based on tender price information (albeit limited to the tenders handled by one firm of quantity surveyors) and is the only historical index explicitly based on tender price data. Of the other indices, no information is available about the calculation of two of them: the Board of Inland Revenue (BIR) index (index 6) and Banister Fletcher's index (index 8). Two others: Redfern (index 4) and CSO CCA (index 5) are, in fact, the same index (though published with different base dates and covering different time periods) and are composites of several other indices of both costs and prices. The remaining indices: Jones (index 1), Saville (index 2), Maiwald (index 3) and the BRS index (index 9) are all cost indices built up on the basis of changes in labour and materials costs, the latter using published measured rates.

With regard to the purposes for which the indices were devised, some were developed for very specific purposes while others were prepared with no particular aim in view. The DOE CNC index was developed, as noted above, as an output price index but suffered from important limitations in this respect. The Redfern and Maiwald indices were developed for use in revaluing long-run series of the value of gross domestic fixed capital formation in buildings and works and in estimating the value of the nation's stock of capital in buildings and works using a perpetual inventory model (see CSO,1985). The CSO CCA index (which, as noted above, is the same as the Redfern index) was issued for current cost accounting purposes. The BIR index was devised for use by Inspectors of Taxes in assessing tax allowances on investment in industrial buildings where figures of initial costs were no longer available. The Jones index was devised as part of a study aimed at assessing changes in efficiency, or real costs, in the London building industry and Saville's index is a direct extension of Jones's. The other indices were developed as general measures of trends, and not, as far as we are aware, for a specific purpose.

9

Historical Cost Price Indices

9.1 JONES'S SELLING PRICE OF BUILDING INDEX

Type of Index

Building cost index.

Series: Coverage and Breakdowns

Series	Table Reference
Selling price of building: 'uncorrected' index	9.1

The index is defined as covering the construction of houses, factories, hospitals and such public works as involve large proportions of brick, stone and woodwork. The construction of harbours, docks, roads and sewers is excluded. Three series were computed by Jones:

1. An 'uncorrected' index.
2. An index corrected for labour and materials.
3. An index corrected for labour and materials and other expenses.

Only one index is reproduced here: the uncorrected index. The other two series represent attempts to calculate changes in 'real' building costs, that is 'the selling cost of a unit of building' corrected for changes in the prices of building materials and labour and other costs. The 'corrected' indices were meant to represent the 'course of selling prices *as it would have been* if the prices of the factors of production had been the same throughout the period, as in the base year', (Jones (1933), p.29). Separate indices for seven building trades (defined below) are also given in the original source.

Base Dates and Period Covered

1910   from 1845 to 1922

Frequency

Annual

Geographical Coverage

London area

Type and Source of Data

Unit rates from *Laxton's Builders' and Contractors' Price Book*.

Method of Compilation

An index is computed for each of seven building trades, each being a weighted average of unit rates. An overall index is then calculated as a weighted average of the seven trade indices. The trades and weights are:

Trade	Weight (%)
Brickwork (including excavator and concreter)	40
Carpentry and joinery	30
Masonry	7
Roofing (tiling and slating)	2
Plumbing	8
Painting	7
Plastering	6

The weights are meant to reflect relative expenditure upon the trades in the period around 1910.

Commentary

Jones referred to his index as the 'selling price of building' but, as it is based upon published unit rates, it is appropriate to define it as a cost, rather than a price, index. The index was later carried forward from 1922 to 1939 by J. Saville (see section 9.2 below).

Publications

(a) Data Source

Jones, G.T. *Increasing Return*, Part II: the London Building Industry, Cambridge: Cambridge University Press, 1933.

(b) Description of Methodology

As above.

Table 9.1 Jones's Selling Price of Building Index, 1845-1922

Base: 1910 = 100

Year	Index	Year	Index	Year	Index
1845	103	1880	110	1915	116
1846	(102)	1881	107	1916	118
1847	(103)	1882	105	1917	140
1848	101	1883	101	1918	166
1849	99	1884	101	1919	273
1850	(96)	1885	100	1920	321
1851	92	1886	98	1921	388
1852	89	1887	97	1922	317
1853	93	1888	96		
1854	100	1889	101		
1855	100	1890	98		
1856	101	1891	95		
1857	100	1892	95		
1858	101	1893	90		
1859	101	1894	90		
1860	100	1895	91		
1861	94	1896	91		
1862	94	1897	93		
1863	99	1898	93		
1864	99	1899	97		
1865	100	1900	100		
1866	104	1901	103		
1867	103	1902	103		
1868	113	1903	103		
1869	104	1904	107		
1870	103	1905	107		
1871	103	1906	106		
1872	103	1907	107		
1873	109	1908	109		
1874	111	1909	107		
1875	108	1910	100		
1876	112	1911	100		
1877	110	1912	101		
1878	110	1913	106		
1879	109	1914	115		

Source: Jones, G.T. (1933), *Increasing Return*, Cambridge: Cambridge University Press.
Note: Figures in brackets are estimates by Jones.

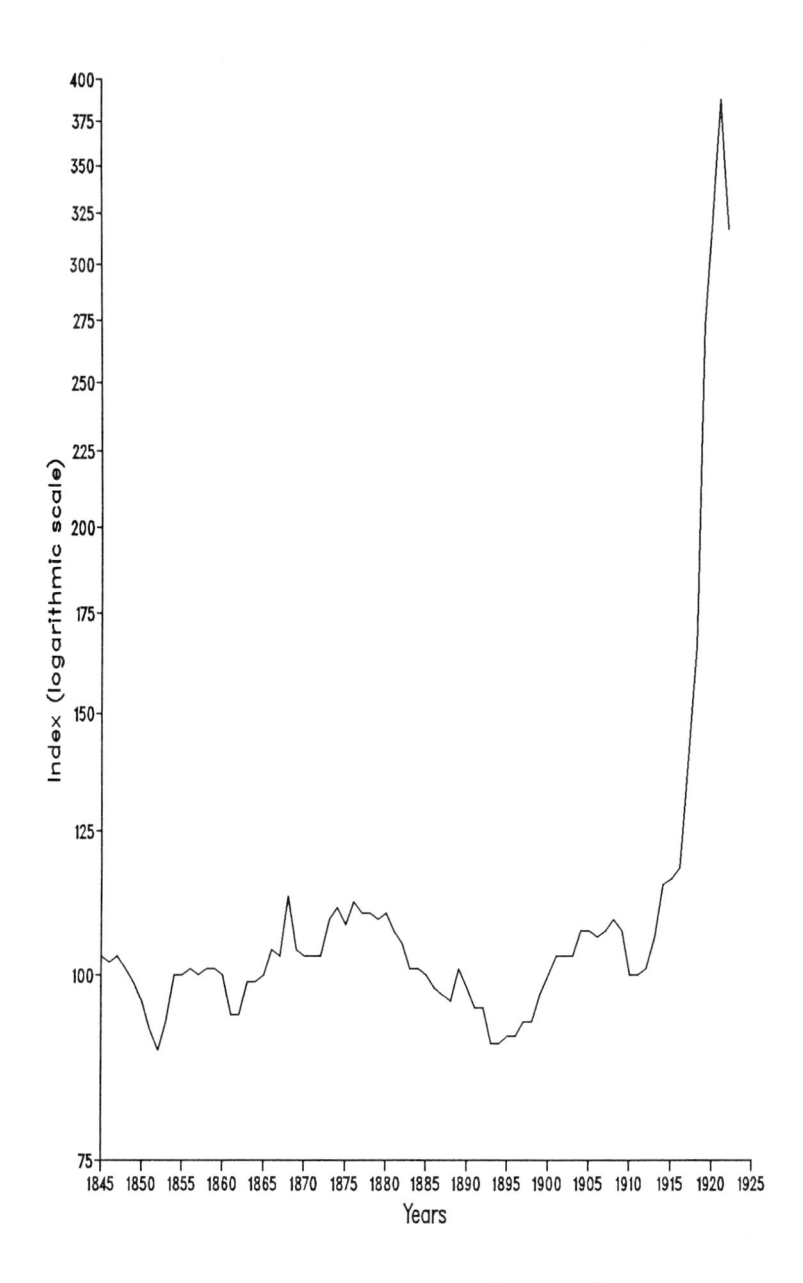

Figure 9.1 Jones's Index of Building Prices (1910 =100), 1845 – 1922

9.2 SAVILLE'S SELLING PRICE OF BUILDING INDEX

Type of Index

Building cost index. This index is a direct extension of Jones's index described in section 9.1.

Series: Coverage and Breakdowns

Series	Table Reference
See section 9.1.	9.2

Base Dates and Period Covered

1910   from 1923 to 1939.

Frequency

Annual.

Geographical Coverage

London area.

Type and Source of Data

See section 9.1.

Method of Compilation

See section 9.1. The only difference between Saville and Jones is that Saville was forced to make certain changes to the unit rates included following revisions in *Laxton's Price Book*. There is, therefore, a slight discontinuity between the two series.

Commentary

See section 9.1.

Publications

(a) Data Source

Saville, J., The measurement of real cost in the London building industry, 1923-1939, *Yorkshire Bulletin of Economic and Social Research*, 1 (2), 1949, pp.67-80.

(b) Description of Methodology

As above.

Table 9.2   Saville's Selling Price of Building Index, 1910 and 1923-1939

Base: 1910 = 100

Year	Index
1910	100
1923	227
1924	221
1925	226
1926	221
1927	225
1928	218
1929	217
1930	216
1931	211
1932	209
1933	199
1934	194
1935	187
1936	184
1937	185
1938	194
1939	207

Source: Saville, J., The measurement of real cost in the London building industry, 1923-1939, *Yorkshire Bulletin of Economic and Social Research*, 1 (2), 1949, pp.67-80.

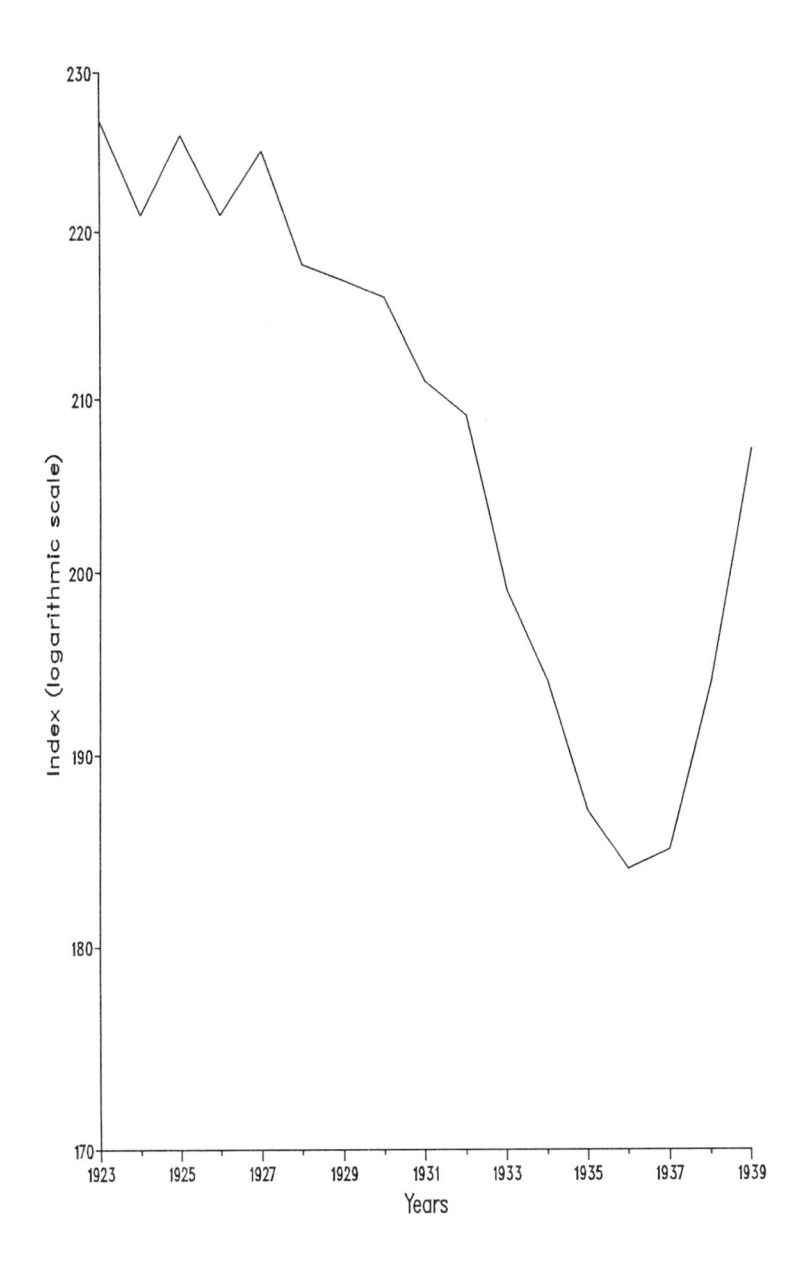

Figure 9.2 Saville's Index of Building Prices (1910 = 100), 1923 – 1939

Spon's handbook of construction cost and price indices

9.3 MAIWALD'S INDICES OF BUILDING COSTS AND OTHER CONSTRUCTION COSTS

Type of Index

Building cost index.

Series: Coverage and Breakdowns

Series	Table Reference
1. Building costs	9.3
2. Other construction costs	9.3

The first index differs from the second in that it includes building materials used in roofing installations and painting, whereas the second index is based upon 'basic materials' only.

Base Dates and Period Covered

1930   from 1845 to 1938.

Frequency

Annual.

Geographical Coverage

United Kingdom.

Type and Source of Data

Based upon wage rates for building craftsmen and labourers and building materials prices. Wage rate information was obtained from official sources and from *Laxton's Builders' and Contractors' Price Book*. Data on materials prices was obtained mainly from *The Builder*, *The Economist*, official sources and *Laxton's Price Book*.

Method of Compilation

Each index is a weighted average of indices for labour and materials prices, each being equally weighted. The materials included in the building costs index are: cement, bricks, stone, wood, iron joists, iron girders, tiles, lead, paint and glass. Each material is weighted equally. The materials included in the 'other construction costs' index are the same except for the last four materials; again each is weighted equally. The labour costs index is based upon wage rates for the following labour: labourers, masons, bricklayers, carpenters and joiners, plasterers, plumbers, painters and slaters; again, each is weighted equally.

Commentary

These indices were calculated as part of work on compiling estimates of long-run domestic capital formation in the United Kingdom extending back to the nineteenth century. The indices are straightforward cost indices, no adjustment being made to allow for changes in technology and their impact on productivity. The choice of materials and labour, on which the indices are based, was dictated largely by the information available. Maiwald comments in particular on the paucity of information available.

Publications

(a) Data Source

Maiwald, K., 'An index of building costs in the United Kingdom, 1845-1938', *Economic History Review*, 2nd series, VII (2), 1954, pp.187-203.

(b) Description of Methodology

As above.

Table 9.3 Maiwald's Indices of Building Costs and Other Construction Costs

Base: 1930 = 100

Year	Building Costs	Other Construction Costs	Year	Building Costs	Other Construction Costs
1845	51.7	50.1	1880	53.7	51.1
1846	53.9	53.6	1881	52.0	49.1
1847	52.5	52.6	1882	52.6	50.1
1848	50.2	50.6	1883	51.4	49.1
1849	48.0	46.7	1884	49.4	47.5
1850	47.2	44.0	1885	49.1	46.8
1851	46.6	43.2	1886	48.0	45.4
1852	47.3	44.6	1887	47.3	45.1
1853	52.9	50.5	1888	47.4	45.2
1854	54.3	50.0	1889	48.9	47.8
1855	53.7	48.4	1890	50.6	50.2
1856	50.7	47.1	1891	49.5	48.4
1857	51.6	47.2	1892	48.9	48.0
1858	50.3	45.4	1893	48.1	46.9
1859	49.8	45.5	1894	47.5	46.6
1860	50.0	46.1	1895	46.9	46.1
1861	49.0	45.4	1896	47.5	47.1
1862	48.6	44.7	1897	48.8	48.0
1863	49.7	45.8	1898	51.1	50.4
1864	50.5	47.4	1899	53.7	53.3
1865	49.7	46.9	1900	56.8	57.7
1866	51.6	48.1	1901	55.8	53.8
1867	50.8	46.9	1902	52.7	50.7
1868	50.1	46.9	1903	51.2	50.8
1869	50.8	47.9	1904	50.2	51.0
1870	51.6	48.8	1905	49.9	49.2
1871	52.4	49.5	1906	51.2	49.7
1872	57.4	55.2	1907	53.0	51.1
1873	63.0	62.2	1908	51.1	50.8
1874	61.0	59.8	1909	50.6	50.2
1875	57.1	53.9	1910	52.0	50.7
1876	55.7	52.5	1911	53.9	51.2
1877	54.8	51.8	1912	56.2	53.5
1878	52.2	49.9	1913	57.8	57.8
1879	50.7	47.8	1914	58.5	58.5

Table 9.3 continued   Base: 1930 = 100

Year	Building Costs	Other Construction Costs
1915	70.4	65.9
1916	82.2	80.4
1917	94.6	93.6
1918	117.5	111.1
1919	146.0	136.6
1920	171.8	166.6
1921	140.5	145.9
1922	113.6	113.7
1923	106.9	106.1
1924	110.9	107.6
1925	111.1	106.9
1926	107.8	105.3
1927	105.9	105.0
1928	101.7	103.2
1929	102.2	102.1
1930	100.0	100.0
1931	96.2	99.0
1932	91.8	94.2
1933	90.0	92.6
1934	90.2	92.5
1935	92.0	93.5
1936	95.6	96.6
1937	100.4	102.6
1938	102.3	108.3

Source: Maiwald, K., An index of building costs in the United Kingdom, 1845-1938, *Economic History Review*, 2nd series, VII (2), 1954, pp.187-203.

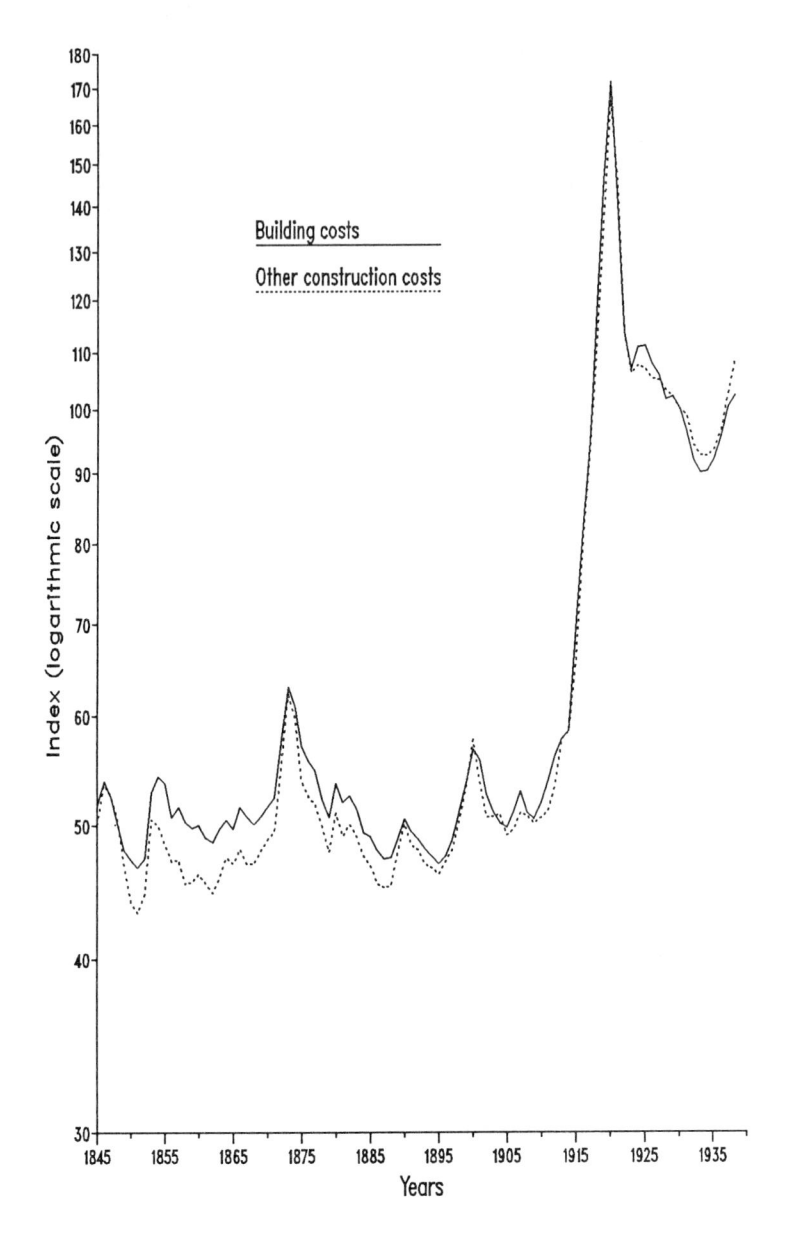

Figure 9.3 Maiwald's Indices of Costs for Building and Other Construction (1930 = 100), 1845 – 1938

9.4 REDFERN'S INDICES FOR HOUSING AND FOR BUILDING AND WORKS

Type of Index

Composite index of building costs and prices.

Series: Coverage and Breakdowns

Series	Table Reference
Buildings and works, 1839/1850-1918	9.4
New housing, 1919-1953	9.4
Other new buildings and works, 1919-1953	9.4

Base Dates and Period Covered

1948   from 1839 to 1953

Frequency

Annual

Geographical Coverage

United Kingdom

Type and Source of Data

The indices are compiled as a composite of other indices of building costs and prices and other information available from a variety of published sources as follows:

1. Jones's index of the selling price of buildings (1845-1922) - see section 9.1.
2. Estimates by H.J. Venning of the building cost of single storey factories and working-class flats (1914, 1924, 1929-1930, 1933) published in *The Architect*, 12 January 1934 and reproduced in an article by Colin Clark in Special Memorandum No. 38 of the London and Cambridge Economic Service, September 1934.
3. Costs per square foot of non-parlour houses for which tenders were received by local authorities in England and Wales (1919-1939) given in *Private Enterprise Housing* (HMSO, 1944).

306   *Spon's handbook of construction cost and price indices*

4. An index of building costs based on measured rates for the period 1920-1933 compiled by Colin Clark and published in London and Cambridge Economic Service, Special Memorandum No. 38.
5. An index of building costs calculated as an average of indices of wage rates and materials prices compiled by *The Economist* for the period 1924-1937 reproduced in Robinson (1939).
6. Estimates of changes in the costs of building a standard three-bedroomed local authority house for the years 1938-39, 1947, 1949 and 1951, given in the Girdwood Reports on *The Cost of Housebuilding*, (HMSO 1948, 1950 and 1952).
7. An index of building costs based on a combination of indices of materials costs, earnings per hour, output per man-hour and overheads for the years 1939 and 1946-1948 given in *The Report of the Working Party on Building* (HMSO, 1950).

The published indices which referred specifically to new housing (i.e. (2), (3) and (6) above) were used in compiling the index of new housing prices covering the years from 1919 onwards. The series listed in items (2), (4), (5), and (7) were brought together, with further details, in Fleming (1966). They are not reproduced here.

Method of Compilation

Redfern describes the methodology as follows: 'These indices have been "spliced together" to form the indices used in this paper. Where, for some period of years, alternative published indices were available and there seemed no strong grounds for regarding one of them as more reliable or relevant than the others, a rough average of the indices has been taken.'

Commentary

The indices were compiled for use in the CSO for converting estimates of the value of the capital stock at current prices to constant prices. For further details see CSO (1985) and *United Kingdom National Accounts* (HMSO annually).

Publications

(a)   Data Source

Philip Redfern, 'Net investment in fixed assets in the United Kingdom, 1938-1953', *Journal of the Royal Statistical Society, Series A (General)*, 118 (2), 1955.

(b)   Description of Methodology

As Above.

Historical cost and price indices 307

Table 9.4 Redfern's Indices for Housing and Building and Works, 1839-1953

Base: 1948 = 100

Year	New Housing	Buildings and Works*	Year	New Housing	Buildings and Works*
1839-50		21	1885		21
1851		20	1886		21
1852		19	1887		21
1853		20	1888		21
1854		21	1889		22
1855		21	1890		21
1856		22	1891		20
1857		21	1892		20
1858		22	1893		19
1859		22	1894		19
1860		21	1895		19
1861		20	1896		19
1862		20	1897		20
1863		21	1898		20
1864		21	1899		21
1865		21	1900		21
1866		22	1901		21
1867		22	1902		20
1868		24	1903		20
1869		22	1904		20
1870		22	1905		20
1871		22	1906		20
1872		22	1907		20
1873		23	1908		20
1874		24	1909		20
1875		23	1910		20
1876		24	1911		20
1877		24	1912		20
1878		24	1913		21
1879		23	1914		21
1880		24	1915		26
1881		23	1916		33
1882		22	1917		42
1883		22	1918		53
1884		22	1919	68	64

Table 9.4 continued Base: 1948 = 100

Year	New Housing	Buildings and Works
1920	78	71
1921	68	58
1922	40	45
1923	41	40
1924	41	42
1925	46	42
1926	46	42
1927	43	42
1928	40	41
1929	38	41
1930	37	40
1931	36	38
1932	34	37
1933	34	35
1934	34	36
1935	33	38
1936	35	40
1937	39	42
1938	38	43
1939	40	42
1940	45	45
1941	50	50
1942	56	56
1943	63	63
1944	71	71
1945	80	80
1946	86	86
1947	94	93
1948	100	100
1949	102	102
1950	105	106
1951	117	123
1952	126	133
1953	126	134

*   including housing up to 1918.
Source: Philip Redfern, Net investment in fixed assets in the United Kingdom, 1938-1953, *Journal of the Royal Statistical Society, Series A (General)*, 118 (2), 1955.

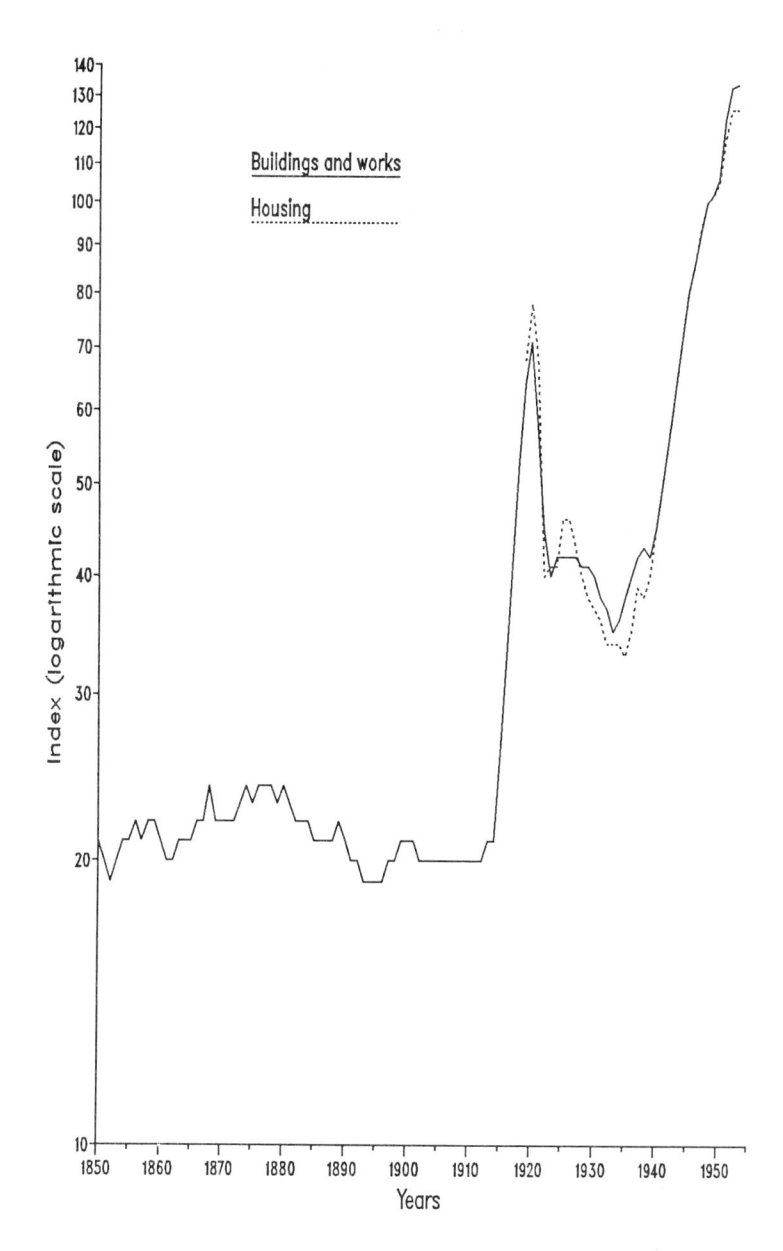

Figure 9.4 Redfern's Indices of Costs for Building and Works and for Housing (1948 = 100), 1850 – 1953

9.5 CSO CCA INDEX FOR NEW BUILDING WORK

Type of Index

Composite index of building costs and prices.

Series: Coverage and Breakdowns

Series	Table Reference
New building work	9.5

Base Dates and Period Covered

1970   from 1888 to 1956

Frequency

Annual

Geographical Coverage

United Kingdom

Type and Source of Data

The series is the same devised by Redfern (given in section 9.4) updated using the DOE 'CNC' index (see section 9.10).

Method of Compilation

See the description in section 9.4.

Commentary

The indices are issued along with a wide variety of other price indices for use in compiling accounts in accordance with the principles of current cost accounting (CCA).

Publications

(a) Data Source

Central Statistical Office, *Price Index Numbers for Current Cost Accounting*, No. 4, April 1977, Table 6 (HMSO, 1977).

(b) Description of Methodology

See section 9.4.

Table 9.5 CSO CCA Index for New Building Work, 1888-1956

Base: 1970 = 100

Year	Index	Year	Index
1888	10	1923	20
1889	11	1924	21
1890	10	1925	21
1891	10	1926	21
1892	10	1927	21
1893	9	1928	20
1894	9	1929	20
1895	9	1930	20
1896	9	1931	19
1897	10	1932	18
1898	10	1933	17
1899	10	1934	18
1900	10	1935	19
1901	10	1936	20
1902	10	1937	21
1903	10	1938	21
1904	10	1939	21
1905	10	1940	22
1906	10	1941	24
1907	10	1942	27
1908	10	1943	31
1909	10	1944	35
1910	10	1945	39
1911	10	1946	42
1912	10	1947	46
1913	10	1948	49
1914	10	1949	51
1915	13	1950	52
1916	16	1951	60
1917	21	1952	65
1918	26	1953	63
1919	31	1954	63
1920	35	1955	67
1921	28	1956	70
1922	22		

Source: Central Statistical Office, *Price Index Numbers for Current Cost Accounting*, No. 4, April 1977, Table 6 (HMSO, 1977).

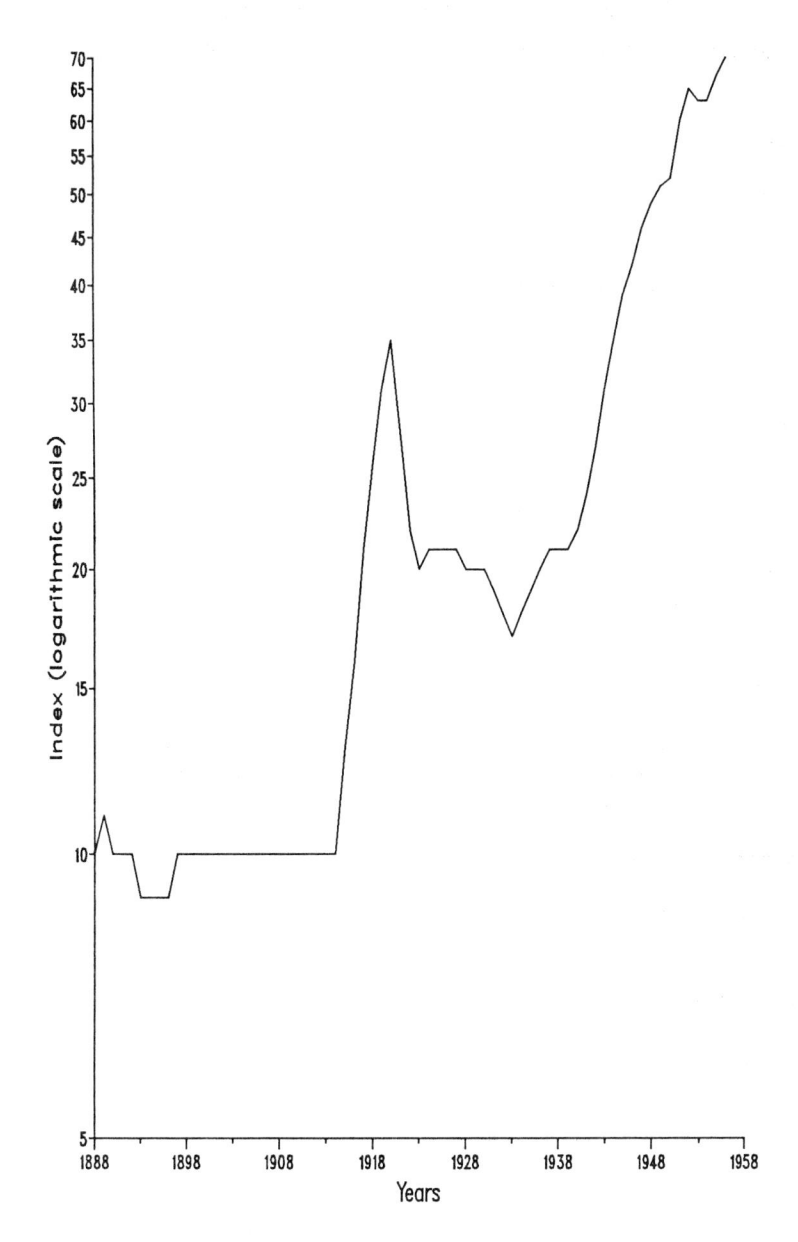

Figure 9.5 CSO CCA Index of Costs for New Building Works (1970 = 100) 1888 – 1956

9.6 BOARD OF INLAND REVENUE (BIR) INDEX OF BUILDING COSTS

Type of Index

Building cost index.

Series: Coverage and Breakdowns

Series	Table Reference
Cost of construction of industrial buildings	9.6

Base Dates and Period Covered

1914　from 1896 to 1955
1949　from 1896 to 1956

Frequency

Annually to 1939 and for irregular dates each year between 1945 and 1956. No series were published for the years 1915-1919 and 1940-1944.

Geographical Coverage

United Kingdom (apparently).

Type and Source of Date

See under 'Method of Compilation' below.

Method of Compilation

For the period 1949-1956, the index is the same as the DOE 'CNC' index – see section 9.10. For the period up to 1949 no official description of the method of compilation and the types and sources of data used was published. The authors are informed, following enquiries, that precise details of sources and methods are no longer available. However, it seems probable that the index was initially compiled retrospectively and that the early series up to the time of the First World War were compiled by the Board of Inland Revenue and that the later series were devised by the quantity surveyors department of the former Ministry of Works. It is also probable that the early series were devised by taking an average of indices of wages and materials costs but whether or not these series related specifically to construction wages and materials is not clear. For materials it is possible that the Sauerbeck index was used - a widely known and used index - although one which covered a range of products including foodstuffs and not merely construction materials - for further details see Devons (1964) p.174-5. For later periods, one may surmise that a variety of secondary sources could have been used at different times including series produced by others, depending on the date of compilation. For example, the Jones index covering the period 1845-1922 (see section 9.1) was first published in 1933. In addition, an index for factory building costs for the period from 1914 was published by Venning in 1934 (*Architect and Building News,* Vol. 137, p.38) and for the period from 1924 an index for building costs (taken as a weighted average of wage rates and materials) was published by *The Economist.* The Venning and *Economist* indices are not reproduced here but for further details see Fleming (1966). All of the indices referred to above received wider publicity by being reproduced in a book on the *Economics of Building* by Robinson in 1939. At the same time, one may also surmise that information obtained by the quantity surveying department of the Ministry of Works as part of its normal activities would have provided data which could be used, either formally or informally, as a basis for estimation.

Commentary

The index was prepared for use by HM Inspectors of Taxes in the settlement of claims to tax allowances where records of the costs of construction of industrial buildings were no longer available. The lack of information about the method of compilation and the sources of data used, makes it difficult to comment on its reliability or usefulness. It gains importance from the fact that it covers a long period of time and is an official index.

Publications

(a) Data Source

Taxation, XXLIII, 1949, p.337 and various issues of *The Chartered Surveyor*, last appearing in the issue for December 1957, p.332.

(b) Description of Methodology

None published.

Table 9.6 Board of Inland Revenue (BIR) Index for New Building Work, 1896-1956

Base: 1949 = 100

Year	Index	Year	Index
1896	19	1930	43
1897	19	1931	39
1898	20	1932	38
1899	21	1933	36
		1934	36
1900	21		
1901	20	1935	37
1902	20	1936	39
1903	20	1937	41
1904	19	1938	42
		1939	41
1905	19		
1906	19	1940	49
1907	20	1941	57
1908	19	1942	60
1909	19	1943	62
		1944	64
1910	19		
1911	19	1945	76
1912	20	1946	85
1913	21	1947	93
1914	22	1948	99
		1949	100
1915	25		
1916	-	1950	101
1917	-	1951	117
1918	-	1952	126
1919	-	1953	123
		1954	123
1920	68		
1921	55	1955	130
1922	40	1956	136
1923	37		
1924	44		
1925	49		
1926	45		
1927	41		
1928	41		
1929	42		

Source: *Taxation*, XXLIII, 1949, p.337 and *The Chartered Surveyor*, December 1957, p.332.

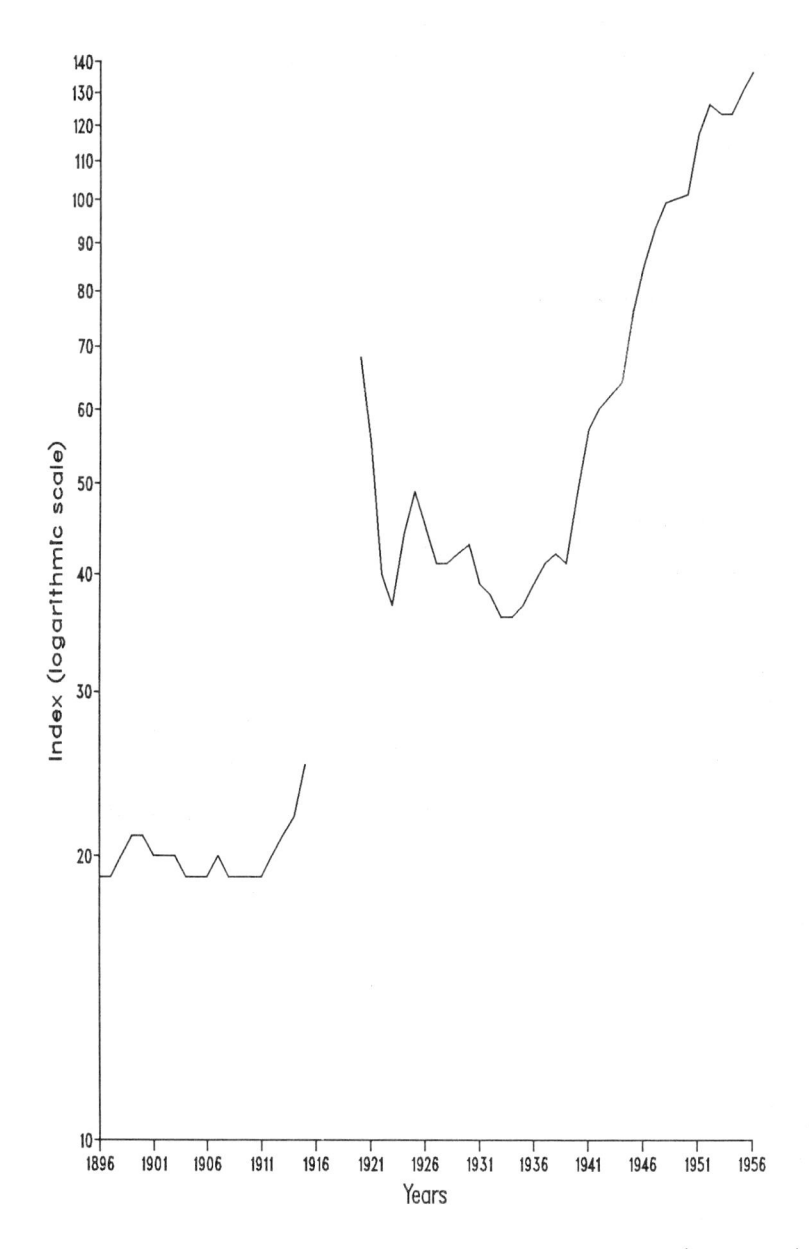

Figure 9.6 Board of Inland Revenue Index of Building Costs (1949 = 100) 1896 – 1956

318 *Spon's handbook of construction cost and price indices*

9.7 VENNING HOPE COST OF BUILDING INDEX

Type of Index

Tender price index.

Series: Coverage and Breakdowns

Series	Table Reference
Cost of building index	9.7

Base Dates and Period Covered

August 1939   from 1914 to June 1975

Frequency

Irregular dates each year. For most of the period covered the series appeared in the journal *Building* (formerly *The Builder*) as a graph only and this has been used as the basis of the annual series reproduced here for the period up to 1964. The subsequent annual series are taken as averages of the successive index numbers published for the month of December each year.

Geographical Coverage

United Kingdom.

Type and Source of Data

Based on tender price information for contracts handled by a firm of quantity surveyors with headquarters in London: Venning Hope Limited (formerly Venning Hope and Partners). Since 1975 Venning Hope Limited have continued to prepare the index for private circulation.

Method of Compilation

The series was not based on a formal statistical methodology (see 'Commentary' below). The series was initiated and compiled (at least for most of the period it covers) by H.J. Venning, FRICS, partner of a firm of quantity surveyors with headquarters in London, using the evidence of work passing through its hands for 'the normal run of building contracts'.

Commentary

The main value of this index arises from the fact that it covers such a long period of time and, unlike many other long-run historical series, is based on actual tender price information. The lack of any formal statistical basis for the calculation of the index is a potentially important limitation in principle but one which it is impossible to assess. Limitation to the contracts handled by one firm of quantity surveyors may also have affected its representativeness and increased the inevitable margins of error to which any index is subject. This limitation may also have affected its representativeness as a national index but, again, this is not possible to assess.

Publications

(a) Data Source

H.J. Venning 'Cost of building chart' published in *Building* (formerly *The Builder*) between 1952 and 1975, last appearing in the issue for 11 July 1975, pp.51-2.

(b) Description of Methodology

None published. The information given above is based on a private communication from H.J. Venning to one of the authors.

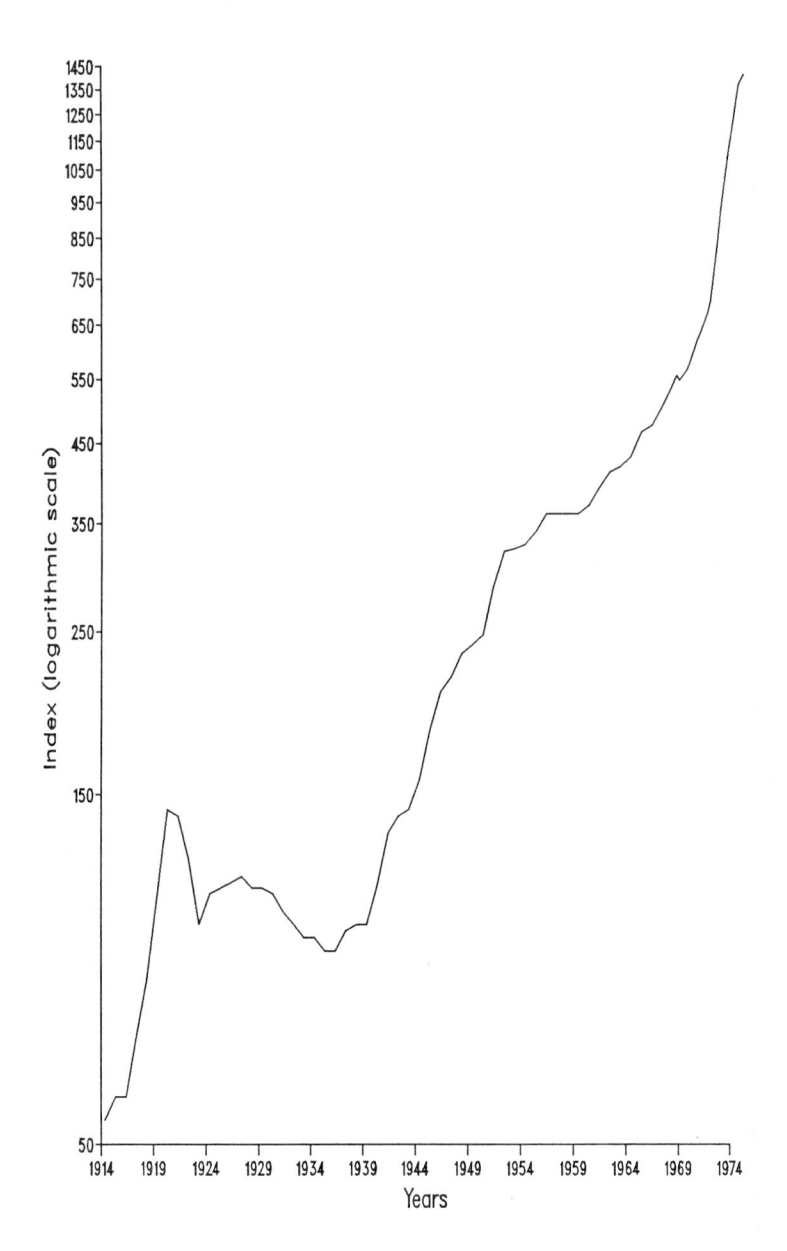

Figure 9.7 Venning Hope Costs of Building Index (August 1939 = 100) 1914 – 1975

9.8 BANISTER FLETCHER'S INDEX OF THE COMPARATIVE COST OF BUILDING

Type of Index

Not known - probably a building cost index.

Series: Coverage and Breakdowns

Series	Table Reference
Comparative cost of building index	9.8

Base Dates and Period Covered

1914   from 1920 to 1946 (except for 1941 and 1942)

Frequency

Annually (except for 1941 and 1942).

Geographical Coverage

Not specified.

Type and Source of Data

Not specified.

Method of Compilation

Not specified.

Commentary

No information is given in the source about the method of compilation or types and sources of data used. It is simply stated that the index '...should be useful in ascertaining the approximate replacement cost for fire insurance or other purpose or the cost of the repetition of a building of which the date of erection and the figures of cost are known'.

Publications

(a) Data Source

Banister Fletcher, *Quantities*, 12th edition, rewritten by A.E. Baylis. Batsford, London, 1947, p.393.

(b) Description of Methodology

None published.

For Table 9.8, see overleaf.

Table 9.8 Banister Fletcher's Comparative Cost of Building Index

Base: 1914 = 100

Year	Index
1914	100
1920	333
1921	270
1922	180
1923	180
1924	200
1925	225
1926	225
1927	200
1928	190
1929	190
1930	185
1931	180
1932	170
1933	158
1934	160
1935	170
1936	180
1937	195
1938	200
1939	200
1942	280
1943	290
1944	305
1945	320
1946	350

Source: Banister Fletcher, *Quantities*, 12th edition. Batsford, London 1947.

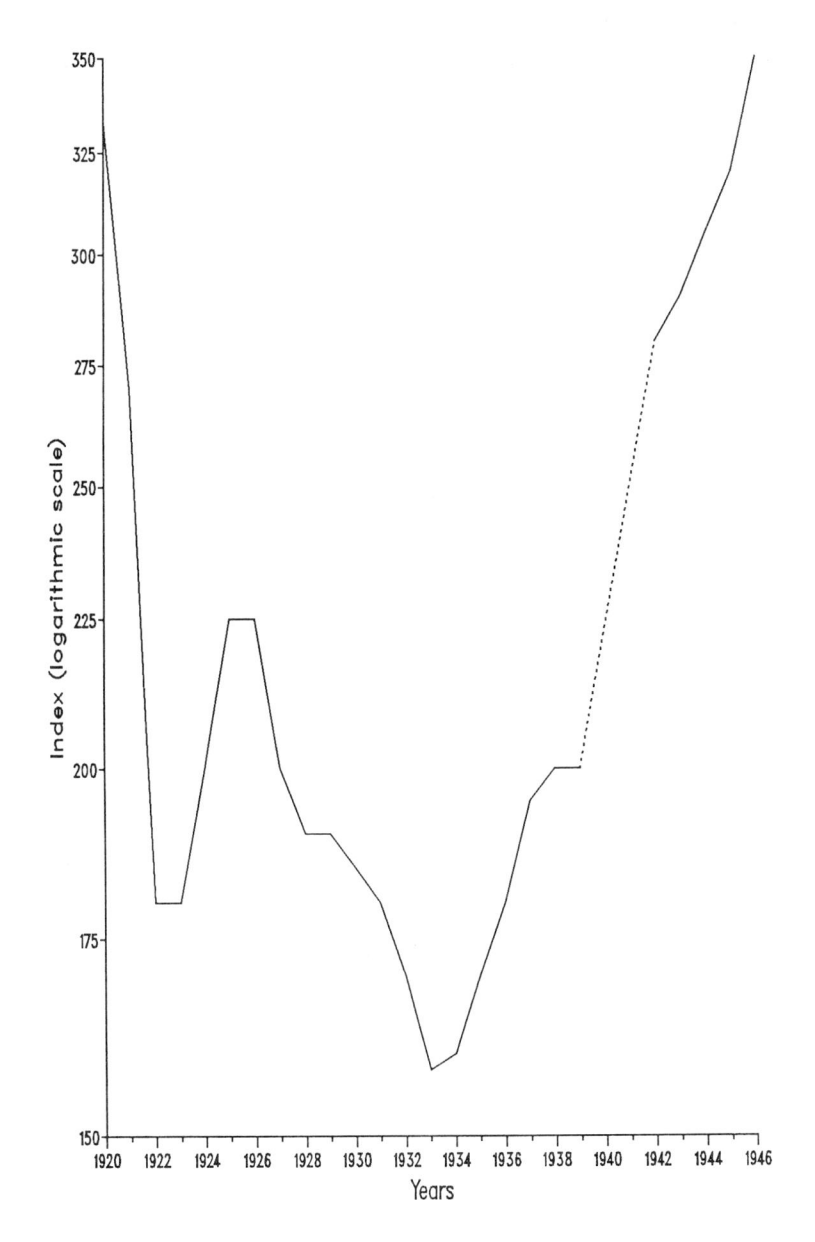

Figure 9.8 Banister Fletcher's Index of the Comparative Cost of Building (1914 = 100), 1920 – 1939

Spon's handbook of construction cost and price indices

9.9 BRS MEASURED WORK INDEX

Type of Index

Building cost index.

Series: Coverage and Breakdowns

Series	Table Reference
Measured work index	9.9

Base Dates and Period Covered

1958-60   from 1939 to 1969 Q2

Frequency

Annually. Quarterly series were prepared but not published except for the period from 1966 - these are included in the data source specified below but not reproduced here.

Geographical Coverage

London area.

Type and Source of Data

Based on unit rates for measured work published in *Architect and Building News* from 1939 to 1948 and in the *Architects' Journal* from 1949 onwards.

Method of Compilation

The index is a weighted average of separate trade indices for the following trades: excavator; drainlayer; concreter; bricklayer; paviour; slater, tiler and roofer; asphalter; carpenter and joiner; iron and steel worker; plasterer; plumber; glazier; painter. These trade indices were published separately in the data source specified below (Table 6.10). Each trade index was based on the movement in the sum of a number of measured rates for each trade. Over time the number of rates included in the index was increased. From 1939 to 1956 it was based on a total of approximately 40 rates. From 1957 to 1960 inclusive the index was based on a total of approximately 280 rates; this number was increased to approximately 370 from the beginning of 1961 and to approximately 500 rates from the beginning of 1963.

Commentary

We describe this index as a building cost index because, although it was based on prices for measured work, these data are based on labour and material costs and make no allowance for changes in market conditions over time (see the discussion in chapter 2).

As a cost index, it may be argued that it had inherent virtues, in principle, over indices based on an average of separate indices for building materials and labour costs in the aggregate, by virtue of the fact that it was built up in detail on the basis of cost movements for the materials and labour used by individual building trades (incorporating thereby an implicit weighting for labour and materials appropriate to each trade). The publication of separate trade indices allowed users to apply alternative patterns of weights to these trade indices to devise an overall index appropriate to their own particular needs. The reliability of the individual trade indices as measures of price movements for those trades was dependent on the number of rates included in their compilation. As noted above, their coverage was expanded over time and greater reliability is attributable therefore to the later series.

Publications

(a) Data Source

Collection of Construction Statistics, Building Research Station, Department of the Environment, 1971.

(b) Description of Methodology

As above.

Table 9.9 BRS Measured Work Index, London Area, 1939 and 1948-1969 Q2

Base: 1958-60 = 100

Year	Index
1939	31
1948	63
1949	65
1950	67
1951	77
1952	87
1953	83
1954	87
1955	89
1956	96
1957	99
1958	99
1959	99
1960	101
1961	106
1962	110
1963	115
1964	121
1965	130
1966	139
1967	147
1968	155
1969 1st Qtr	160
2nd Qtr	163

Note: The annual value is taken as the value at the second quarter up to and including 1954 and as the average of the four quarters thereafter.

Source: DOE, Building Research Station, *Collection of Construction Statistics*, 1971.

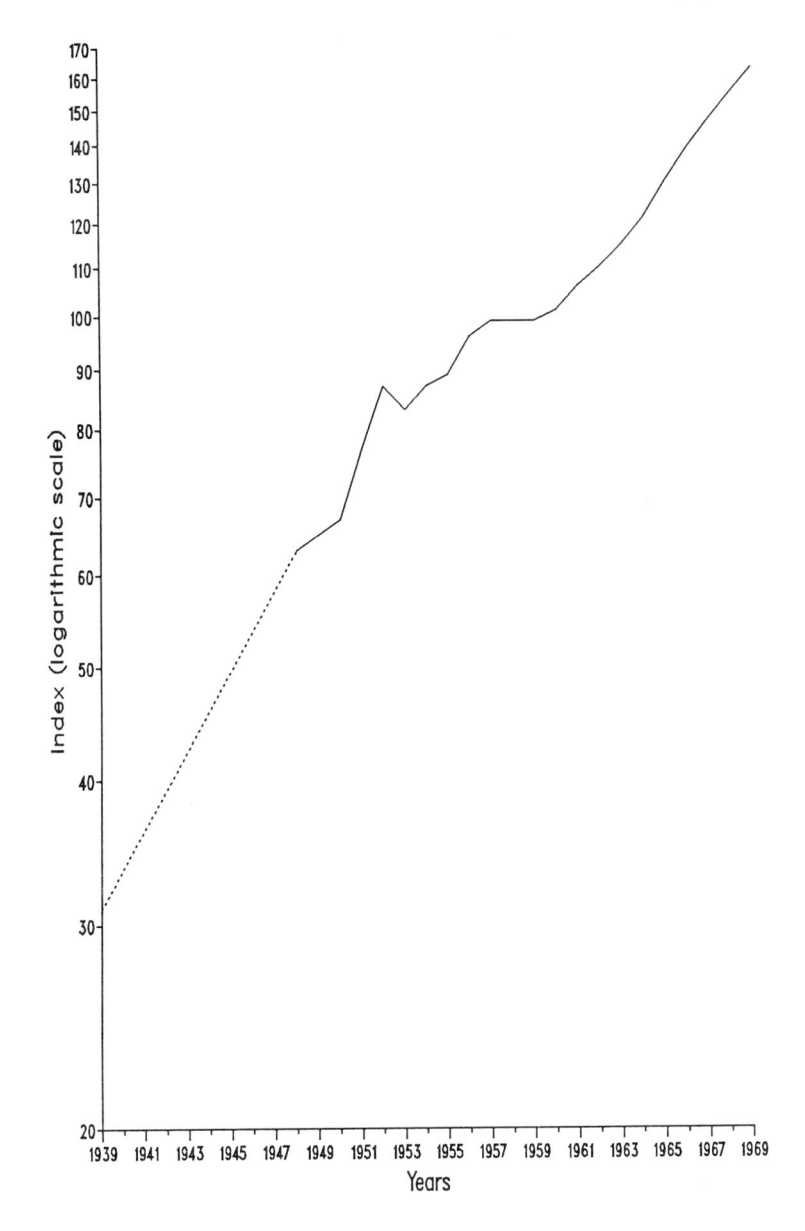

Figure 9.9 BRS Measured Work Index, London Area (1958 – 60 = 100) 1939 – 1969 Q2

332 *Spon's handbook of construction cost and price indices*

9.10 DOE COST OF NEW CONSTRUCTION (CNC) INDEX

Type of Index

Building cost index.

Series: Coverage and Breakdowns

Series	Table Reference
Cost of new construction	9.10

Separate indices for 'new housing' and 'other new work', on which the CNC index itself was based until 1969 were published unofficially, along with two other related series, for part of the period (see below) but they are not reproduced here.

Base Dates and Period Covered

1949	from 1949 to 1954
1954	from 1954 to 1963
1963	from 1963 to 1970
1970	from 1970 to 1980 Q1

The pre-1970 series are shown with 1970 = 100 in the table below. These were converted arithmetically by the authors.

Frequency

Quarterly.

Geographical Coverage

United Kingdom.

Type and Source of Data

Based on official indices of labour and material costs and estimated changes in productivity, overheads and profits.

Method of Compilation

The CNC index was of the composite type based on labour and materials cost changes plus an allowance for changes in productivity, overheads and profits. It was computed as a base-weighted average of separate indices for 'new housing' and 'other new work'. In turn, each of these indices was calculated as a base-weighted average of an index of labour costs adjusted for productivity change (in order to give an index of labour costs per unit of output) and an index of materials prices for housebuilding or for other construction as appropriate. The index of labour costs used was based on official statistics of average hourly earnings (as opposed to wage rates) in the industry. For materials costs the official indices of the prices of 'housebuilding materials' and 'construction materials' produced as part of the official series of 'wholesale price indices' compiled by the Board of Trade and its successor departments. The adjustment for productivity change was made using an index of estimated materials input per man at constant prices. With regard to changes in overheads and profits, there were no satisfactory statistics and in practice a percentage mark-up was taken, the percentage being revised periodically in the light of information obtained from an annual census of production and from inland revenue taxation statistics.

The last official description of the index (with base 1970 = 100) published in *Housing and Construction Statistics: Notes and Definitions Supplement, 1977* (HMSO, 1977) was as follows:

'The index provides a measure of changes in the cost to clients of new building and civil engineering work. It is designed to measure general cost changes and not changes in the cost of any specific type of new work. The index relates to the cost of construction completed in a given quarter and does not monitor the changes in the levels of tenders. The work done in a period is, of course, made up of orders let at varying tender price levels in previous time periods. The index takes account of movements in material prices and in labour costs, making allowance for varying levels of profits, etc., and is calculated according to the formula:

$$\frac{\left[\dfrac{M_o I_m}{100} + \dfrac{E_o I_e}{I_{pr}} \right] \times 100}{1 - C_t}$$

M_o and E_o are the base-year (1968) proportions of gross output accounted for by (a) the value of materials used and (b) the labour costs defined as earnings, employers' social security payments, and other relevant employers' liabilities: they are derived from the 1968 census of production.

C_t is the estimated proportion in period 't' of gross output accounted for by overheads (including plant hire charges) and profits. The value in the base-year (C_o) was obtained from the census of production, and C_t is produced by slightly adjusting C_o for each year thereafter.

I_m is an index of the price of building materials used in construction.

I_e is an index of labour costs per operative, based on the statistics of all employees' earnings compiled by the Department of Employment.

I_{pr} is a constant price index of materials input per worker. It is obtained by, first, subtracting labour costs, profits and overheads per man from the collected figures of gross new work output divided by numbers of men employed on this work, and then dividing the resultant estimate of materials, at current prices used per man, by the price index (I_m) of building materials. A four-quarter moving average of I_{pr} is used as an index of productivity, to adjust the index I_e of labour costs.

All indices are expressed for the formula in terms of base-year (1968) = 100, and converted to 1970 = 100.'

Commentary

The CNC index was compiled for use in converting statistics of the value of construction output and also statistics of expenditure on gross domestic fixed capital formation in buildings and works to constant prices. It was meant to measure changes, therefore, not in tender prices for future work but changes in *current costs* (to the client) of new work (as opposed to repairs and maintenance work) as it is executed in each successive time period. In effect, therefore, in any particular time period such an index should reflect the movement of an amalgam of prices for various *parts* of contracts passing through the construction process in that period. These prices, of course, will have been set at various periods in the past depending on the length of the construction process (although in some cases they will have been adjusted in line with movements in labour and/or materials costs since the tender date under variation of price clauses in the contracts). However, the index suffered from several limitations arising from the methodology and the statistical information on which it was based. It was eventually accepted as being unreliable and replaced by the DOE output price indices (given in chapter 4). For a more detailed critique of the methodology and its limitations see Fleming (1980), pp.267-70.

As indicated above, sub-indices of the CNC index for 'new housing' and 'other new work' were published unofficially together with an index for repair and maintenance work (devised in the same way as the CNC index except that it was assumed that productivity on repair and maintenance work remained unchanged over time) and an index for 'all building' obtained by combining the CNC index and the repairs and maintenance index. These were published in a series of articles entitled 'Productivity and Prices' prepared by C.F. Carter in the journal *Building* (formerly *The Builder*), first appearing in the issue for 15 July 1955, pp.38-9.

Publications

(a) Data Source

1949-1962: *Monthly Bulletin of Construction Statistics*, January 1968, (Ministry of Public Building and Works)
1963-1980: *Housing and Construction Statistics*, (HMSO).

(b) Description of Methodology

'New index of the cost of building and civil engineering works', *Board of Trade Journal*, 170, 12 May 1956, pp.608-9. Subsequent descriptions appeared in *Housing and Construction Statistics: Notes and Definitions Supplement*, (HMSO, annually).

Table 9.10 DOE Cost of New Construction ('CNC') Index, 1949-1980 Q1

Base: 1970 = 100*

Year	Q1	Q2	Q3	Q4	Average
1949	50	49	49	49	49
1950	49	49	50	51	50
1951	53	58	60	60	58
1952	62	63	62	62	62
1953	62	61	61	60	61
1954	60	61	61	62	61
1955	62	65	65	65	64
1956	65	67	67	67	67
1957	68	68	70	70	69
1958	70	70	70	69	70
1959	68	68	68	68	68
1960	68	69	69	69	69
1961	70	71	71	73	72
1962	73	74	74	75	74
1963	76	76	77	77	76
1964	77	78	79	79	78
1965	79	81	82	82	81
1966	83	85	86	86	85
1967	86	86	87	87	86
1968	89	90	92	92	90
1969	93	94	95	95	94
1970	97	100	101	103	100
1971	103	105	109	111	107
1972	112	115	119	132	119
1973	135	141	151	161	147
1974	170	180	190	200	185
1975	210	220	228	236	224
1976	238	246	264	281	257
1977	290	298	304	314	301
1978	314	319	330	343	326
1979	348	370	394	412	381
1980	425				

* Indices from 1949 to 1970 inclusive converted arithmetically by the authors from bases of 1949 = 100 (1949-54), 1954 = 100 (1954-63) and 1963 = 100 (1963-70) to 1970 = 100.

Source: Ministry of Public Building and Works, *Monthly Bulletin of Construction Statistics*, January 1968 and HMSO, *Housing and Construction Statistics*.

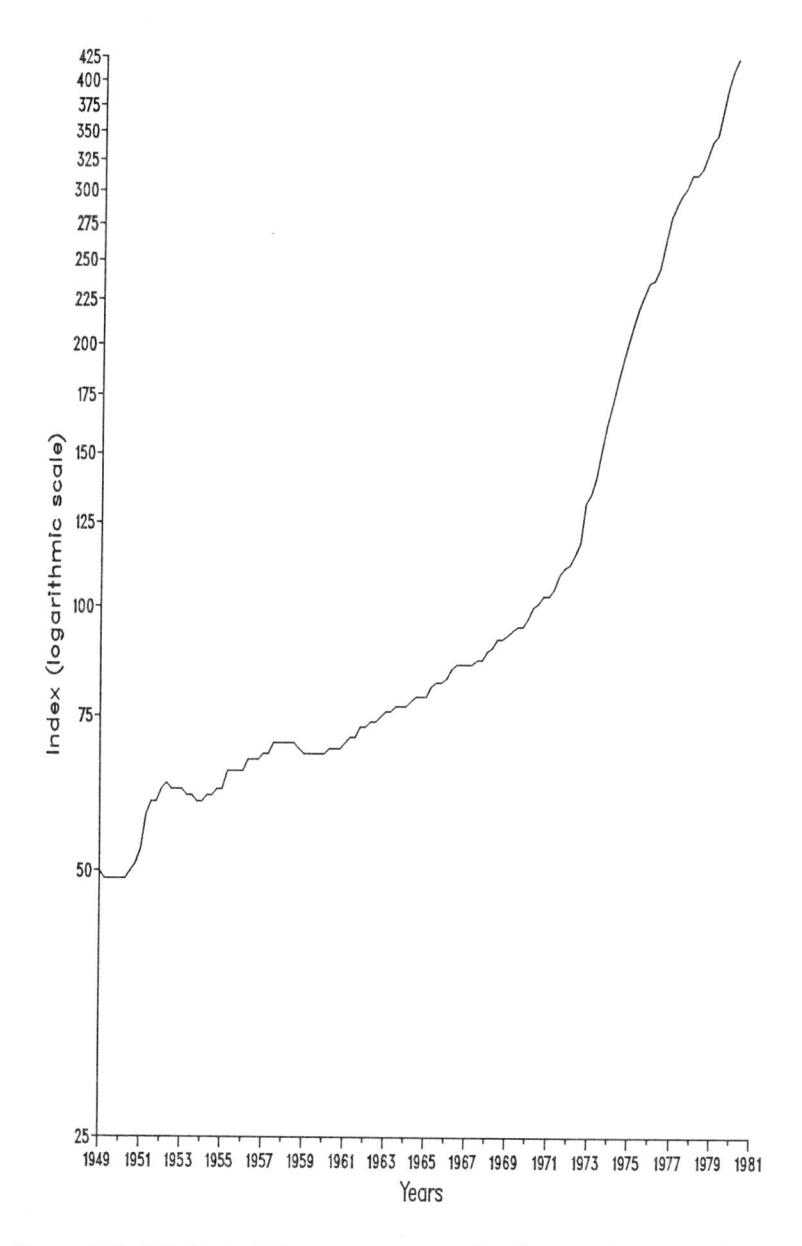

Figure 9.10 DOE Cost of New Construction (CNC) Index (1970 = 100)
1949 Q1 – 1980 Q1

10

Comparison and Review of Historical Indices

To compare the long-run historical indices it is necessary to rescale them to a common base. As the choice of a single year for the base tends to distort the comparison here by attributing a reliability to year-to-year differences between one index and another which is unmerited, a base spanning several years (1924-1933) has been taken; thus for each series the average of the series for the years from 1924 to 1933 inclusive equals 100. All of the indices, except two, are included in the comparison, the two excluded being the BRS measured work index and the DOE CNC index which commence in 1939 and 1949 respectively. Between them, the remaining indices allow a comparison for the period from 1845 to 1974. The rebased index numbers are given in Table 10.1 and a graph of the series in Figure 10.1. It is not possible to show all of the series very clearly on a single graph but it should be appreciated that the purpose is merely to give a broad indication of trends. However, to aid clarity, two of the indices (Maiwald's 'other construction' index and the Banister Fletcher index) are omitted from the graph.

In comparing the various *current* construction indices considered in Part B of this book, attention is devoted to differences between one type of index and another, as well as reviewing the evidence they provide about the magnitude of cost and price trends over time. We concentrate here, however, on reviewing the evidence which the historical indices provide about long-run trends, rather than comparing one index with another, because the available series do not allow close distinctions to be drawn between historical *cost* indices on the one hand and *price* indices on the other.

Given the nature of the series available, it is to be expected that there should be some disagreement about the magnitude of changes from year to year and about the precise timing of turning points. Differences of this sort, however, are generally fairly small. What is more striking is the fact that, viewing the period as a whole and taking the evidence at its face value (bearing in mind that some of the indices are not independent of the others), there is a close measure of agreement about trends.

The salient feature of the period as a whole is that it falls into three distinct parts separated by the two world wars. Before the First World War the experience appears to have been one of relative stability in prices. The inter-war period was characterised at first by rapid inflation (initiated during the war itself) followed by rapid deflation, although prices did not fall as much as they had previously risen. Inflation then set in again with the outbreak of the Second World War (indeed somewhat earlier) but, in contrast to the experience following the First World War, continued to rise at a rapid rate through the whole of the succeeding period. We now turn to the magnitude of the changes before comparing them with changes in the economy as a whole.

Prices reached a peak in the inter-war period in 1920-1921 around three to three and a half times their immediate pre-war level and around four times their 1845 level. Prices then fell but at their trough in the 1930s remained at no less than 50-75 per cent higher than in 1914 and by 1939 had gone back up to roughly double their 1914 level. As prices between 1845 and 1914 had increased very little, this meant that by the outbreak of the Second World War in 1939 prices were also roughly double their 1845 level. During the war prices almost doubled again, so that by the end of the war in 1945 the construction price level stood at around four times its 1914 level.

After the Second World War, building costs and prices continued to rise strongly and by 1949 had risen to around four and a half times their 1914 level. Marrying together the evidence from the historical indices surveyed in this chapter with the evidence from the current indices surveyed earlier in chapter 7, one can say that between 1949 and 1989 building costs and prices both rose by around 17-fold (although in the intervening years cost and price movements diverged from time to time). Combining this evidence with that for the earlier period back to 1914, discussed above, we may say that over the long run of 75 years from 1914 to 1989, it appears that building costs and prices increased by around 80 times.

Turning, finally, to compare this evidence for construction with price trends in the economy as a whole, the most striking feature is the fact that construction costs show a very strong tendency, maintained over a long period, to rise much more rapidly than prices in general. The official general price indices, which form the basis of this comparison, extend back to 1914 and are given in Appendix A. Over the post-Second World War period from 1949 to 1989, it appears that prices in general increased by around 14 times (see Appendix tables A2-A4 which show increases of 14.2 for retail prices, 14.3 for total home costs and 13.5 for capital goods prices respectively). This compares with the 17-fold increase in building costs and prices discussed above. Over the longer period from 1914 to 1989, combining the evidence from the official 'cost of living' index from 1914 to 1947 (Appendix table A1) with that from the subsequent retail prices index (Appendix table A2), it appears that prices in general rose by around 32 times as against the corresponding figure of 80 times for building costs and prices.

In conclusion we would emphasise that these comparisons are simply an attempt to summarise what the wide range of information presented in this book shows about long-run changes. Care must be taken in interpreting the figures. It must be remembered that building *cost* indices, in particular, generally make no allowance for productivity changes in the construction industry. Further, over the long period surveyed here, higher building prices have resulted from higher building specifications, particularly with regard to the engineering services required in modern buildings, and, on this account, the long-run indices face a difficulty in maintaining a standard of comparison. These problems afflict the long-term historical evidence in particular. They are much less acute for the current indices because of the greater volume of information available, especially with regard to tender prices, and improved methods of calculation.

For Table 10.1, see overleaf.

342 *Spon's handbook of construction cost and price indices*

Table 10.1 Comparison of Selected Long-run Historical Indices, 1845–1974

Base: 1924-33 = 100*

Year	Jones & Saville**	Maiwald: Buildings	Maiwald: Other construction	Redfern Buildings & works	CSO CCA	BIR	Venning Hope	Banister Fletcher
1845	48	51	49					
1846	47	53	53					
1847	48	52	52					
1848	47	49	50					
1849	46	47	46					
1850	44	46	43	53				
1851	43	46	43	50				
1852	41	46	44	48				
1853	43	52	50	50				
1854	46	53	49	53				
1855	46	53	48	53				
1856	47	50	46	55				
1857	46	51	46	53				
1858	47	49	45	55				
1859	47	49	45	55				
1860	46	49	45	53				
1861	43	48	45	50				
1862	43	48	44	50				
1863	46	49	45	53				
1864	46	50	47	53				

Comparison and review of historical indices 343

1865	46	49	46		53
1866	48	51	47		55
1867	48	50	46		55
1868	52	49	46		60
1869	48	50	47		55
1870	48	51	48		55
1871	48	51	49		55
1872	48	56	54		55
1873	50	62	61		58
1874	51	60	59		60
1875	50	56	53		58
1876	52	55	52		60
1877	51	54	51		60
1878	51	51	49		60
1879	50	50	47		58
1880	51	53	50		60
1881	49	51	48		58
1882	49	52	49		55
1883	47	51	48		55
1884	47	49	47		55
1885	46	48	46		53
1886	45	47	45		53
1887	45	46	44		53
1888	44	47	44	51	53
1889	47	48	47	56	55
1890	45	50	49	51	53

Table 10.1 continued

Year	Jones & Saville**	Maiwald: Buildings	Maiwald: Other construction	Redfern Buildings & works	CSO CCA	BIR	Venning Hope	Banister Fletcher
1891	44	49	48	50	51			
1892	44	48	47	50	51			
1893	42	47	46	48	45			
1894	42	47	46	48	45			
1895	42	46	45	48	45			
1896	42	47	46	48	45	45		
1897	43	48	47	50	51	45		
1898	43	50	50	50	51	48		
1899	45	53	52	53	51	50		
1900	46	56	57	53	51	50		
1901	48	55	53	53	51	48		
1902	48	52	50	50	51	48		
1903	48	50	50	50	51	48		
1904	49	49	50	50	51	45		
1905	49	49	48	50	51	45		
1906	49	50	49	50	51	45		
1907	49	52	50	50	51	48		
1908	50	50	50	50	51	45		
1909	49	50	49	50	51	45		

1910	46	51	50	50	51		45		
1911	46	53	50	50	51		45		
1912	47	55	53	50	51		48		
1913	49	57	57	53	51		50		
1914	53	57	58	53	51		53		52
1915	54	69	65	65	66				
1916	55	81	79	83	81				
1917	65	93	92	105	106		60		
1918	77	115	109	133	131			50	
1919	126	143	134	160	157			53	
1920	148	169	164	178	177	163		53	173
1921	179	138	144	145	141	132		64	140
1922	147+	112	112	113	111	96		77	94
1923	105	105	104	100	101	89		101	94
1924	102	109	106	105	106	105			104
1925	104	109	105	105	106	117	117	103	117
1926	102	106	104	105	106	108	117	105	117
1927	104	104	103	105	106	98	107	104	
1928	101	100	102	103	101	98	103	99	
1929	100	100	100	103	101	100	103	99	
1930	100	98	98	100	101	103	101	96	
1931	98	95	97	95	96	93	96	94	
1932	97	90	93	93	91	91	92	88	
1933	92	88	91	88	86	86	88	82	
1934	90	89	91	90	91	86	88	83	

Table 10.1 continued

Year	Jones & Saville**	Maiwald: Buildings	Maiwald: Other construction	Redfern Buildings & works	CSO CCA	BIR	Venning Hope	Banister Fletcher
1935	86	90	92	95	96	89	85	88
1936	85	94	95	100	101	93	85	94
1937	86	99	101	105	106	98	90	101
1938	90	101	107	108	106	100	92	104
1939	96			105	106	98	92	104
1940				113	111	117	104	
1941				125	121	136	122	
1942				140	136	144	129	146
1943				158	157	148	132	151
1944				178	177	153	145	159
1945				200	197	182	169	166
1946				215	212	203	191	166
1947				233	232	222	200	
1948				250	247	237	215	
1949				255	258	239	221	
1950				265	263	242	227	
1951				308	303	280	264	
1952				333	328	301	295	
1953				335	318	294	297	
1954					318	294	301	

Year			
1955		338	313
1956		354	331
1957			331
1958			331
1959			331
1960			341
1961			359
1962			378
1963			384
1964			396
1965			428
1966			437
1967			463
1968			497
1969			515
1970			544
1971			594
1972			701
1973			898
1974			1135

* Indices have been converted arithmetically by the authors to a base of 1924-33 = 100 from their various original bases as shown in Tables 9.1-9.8. Differences between Redfern and CSO CCA are due to rounding arising from rebasing.

** Saville's index (1923-39) is an extension of Jones's (1845-1922) and has been used for the purpose of rebasing Jones's index on the years 1924-33. The two are linked on the year 1910.

+ Some discontinuity in the series at this point.

Source: see Tables 9.1-9.8.

Comparison and review of historical indices   347

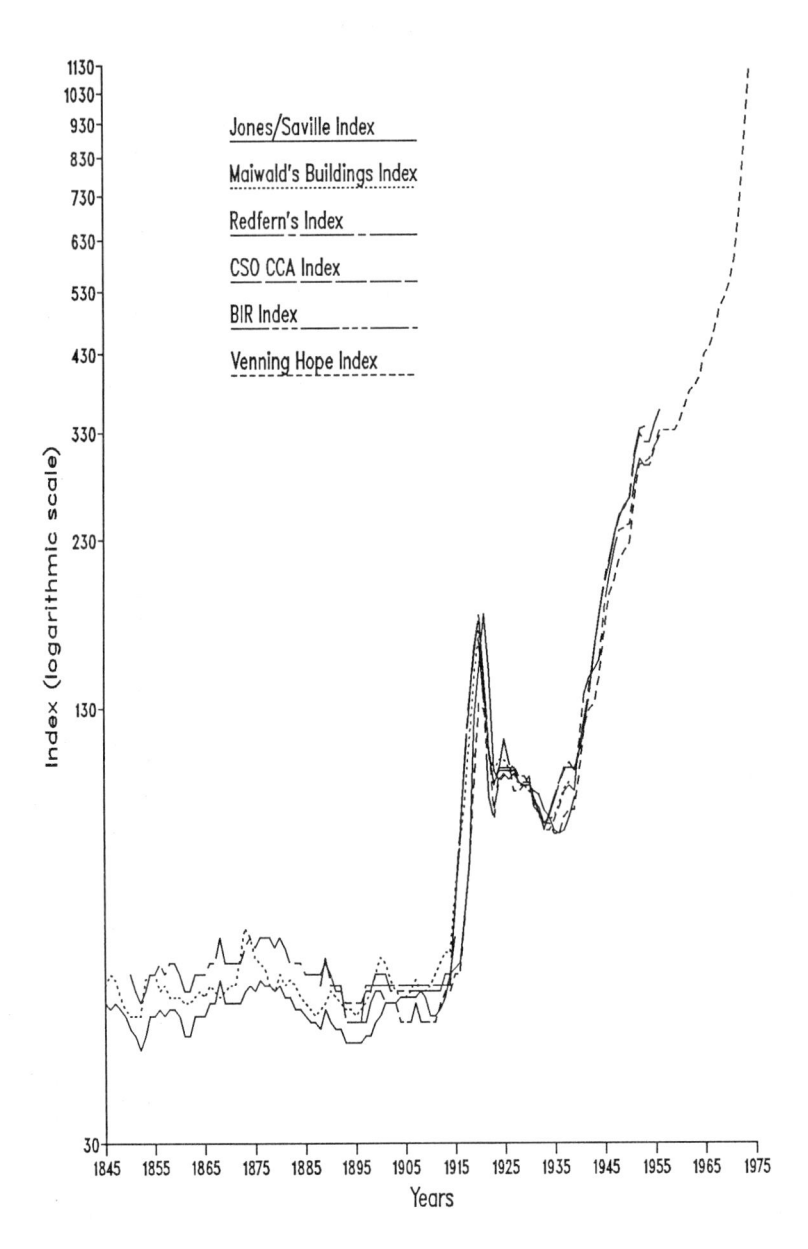

Figure 10.1 Comparison of Selected Long-run Historical Indices (1924 – 1933 = 100), 1845 – 1974

Bibliography

Central Statistical Office (1985), *United Kingdom National Accounts, Sources and Methods*, 3rd edition. HMSO, London.

Davis, Belfield and Everest (1975), *Spon's Architects' and Builders' Price Book*, 100th edition, E. & F.N. Spon, London.

Devons, E. (1964), *An Introduction to British Economic Statistics*. University Press, Cambridge.

Fleming, M.C. (1965), Costs and prices in the Northern Ireland construction industry, *Journal of Industrial Economics*, XIV (1), pp.42-54.

Fleming, M.C. (1966), The long-term measurement of construction costs in the United Kingdom, *Journal of the Royal Statistical Society, Series A (General)*, 129 (4), pp.534-556 and Correction, Vol. 130, Part 2, 1967, p.282.

Fleming, M.C. (1980), *Construction and the Related Professions*, Reviews of UK Statistical Sources Vol. 12. Pergamon Press, Oxford.

Fleming, M.C. (1986), *Spon's Guide to Housing, Construction and Property Market Statistics*. E. & F.N. Spon, London.

Fleming, M.C. and Nellis, J.G. (1984), *The Halifax House Price Index: Technical Details*. Halifax Building Society, Halifax.

Fleming, M.C. and Nellis, J.G. (1985), The application of hedonic indexing methods: a study of house prices in the United Kingdom, *Statistical Journal of the United Nations Economic Commission for Europe*, 3 (3), pp.249-70.

Fleming, M.C. and Nellis, J.G. (1987), *Spon's House Price Data Book*. E. & F.N. Spon, London.

Fleming, M.C. and Nellis, J.G. (1989), *The New Nationwide Anglia House Price Index: A Technical Guide*. Nationwide Anglia Building Society, London.

Fleming, M.C. and Nellis, J.G. (1991), *Essence of Statistics for Business* (forthcoming, Philip Allan).

Jones, G.T. (1933), *Increasing Return*. University Press, Cambridge.

Robinson, H.W. (1939), *The Economics of Building*. King, London.

Tysoe, Brian A. (1981), *Construction Cost and Price Indices: Description and Use*, E. & F.N. Spon, London.

Appendix A

General Indices of Prices

List of Tables

A1   Cost of Living Index, 1914-1948

A2   Index of Retail Prices, 1948-1989

A3   Index of Total Home Costs, 1948-1989

A4   Index of Capital Goods Prices, 1948-1989

Table A1 Cost of Living Index, 1914-1948* (June each year)

Base: July 1914 = 100

Year	Index	Year	Index
1915	125	1935	140
1916	145	1936	144
1917	175-180	1937	152
1918	200	1938	155
1919	205	1939	153
1920	250	1940	181
1921	219	1941	200
1922	180	1942	199
1923	169	1943	198
1924	169	1944	200
1925	172	1945	204
1926	168	1946	203
1927	163	1947	203
1928	165	1948 June	223
1929	160		
		1948 Average	219
1930	154		
1931	145		
1932	142		
1933	136		
1934	138		

* Official 'Cost of Living' index numbers from 1914 to June 1947 (the termination date of the index). The index has been carried forward to 1948 to provide a link with Table A2, which gives the 'index of retail prices' (RPI) by which it was replaced in 1947, using the movement in that index from June 1947.

Source: Department of Employment *Retail Prices Indices, 1914-1986*, HMSO, London, 1987.

General indices of costs and prices

Table A2 Index of Retail Prices, 1948-1989

Base: 1985 = 100

Year	Index	Year	Index
1948	8.3	1970	19.6
1949	8.6	1971	21.4
		1972	23.0
1950	8.8	1973	25.1
1951	9.7	1974	29.1
1952	10.5		
1953	10.9	1975	36.1
1954	11.1	1976	42.1
		1977	48.8
1955	11.6	1978	52.8
1956	12.1	1979	59.9
1957	12.6		
1958	13.0	1980	70.7
1959	13.0	1981	79.1
		1982	85.9
1960	13.2	1983	89.8
1961	13.6	1984	94.3
1962	14.2		
1963	14.5	1985	100.0
1964	15.0	1986	103.4
		1987	107.7
1965	15.7	1988	113.0
1966	16.3	1989	121.8
1967	16.7		
1968	17.5	1990	
1969	18.4	1991	
		1992	
		1993	
		1994	

Source: *Economic Trends Annual Supplement* (HMSO).

Table A3   Index of Total Home Costs, 1948-1989*

Base: 1985 = 100

Year	Index	Year	Index
1948	8.4	1970	19.1
1949	8.6	1971	21.2
		1972	23.3
1950	8.6	1973	25.2
1951	9.4	1974	29.4
1952	10.1		
1953	10.5	1975	37.5
1954	10.7	1976	43.0
		1977	48.3
1955	11.1	1978	54.1
1956	11.8	1979	61.0
1957	12.2		
1958	12.8	1980	72.3
1959	13.0	1981	79.5
		1982	85.2
1960	13.3	1983	90.0
1961	13.7	1984	94.9
1962	14.1		
1963	14.4	1985	100.0
1964	14.9	1986	102.6
		1987	107.8
1965	15.5	1988	114.8
1966	16.1	1989	123.1
1967	16.6		
1968	17.1	1990	
1969	17.7	1991	
		1992	
		1993	
		1994	

* Implied gross domestic product deflator at factor cost obtained by dividing current price expenditure-based estimates of gross domestic product by the corresponding estimates at 1985 prices.

Source: *Economic Trends Annual Supplement* (HMSO).

Table A4 Index of Capital Goods Prices, 1948-1989*

Base: 1985 = 100

Year	Index	Year	Index
1948	9.1	1970	18.9
1949	9.2	1971	20.7
		1972	22.8
1950	9.4	1973	26.4
1951	10.4	1974	32.1
1952	11.6		
1953	11.7	1975	39.4
1954	11.7	1976	45.1
		1977	50.7
1955	12.2	1978	56.6
1956	12.8	1979	65.4
1957	13.3		
1958	13.6	1980	77.8
1959	13.5	1981	85.5
		1982	88.0
1960	13.6	1983	90.9
1961	13.9	1984	94.7
1962	14.3		
1963	14.8	1985	100.0
1964	15.1	1986	104.4
		1987	109.8
1965	15.5	1988	116.4
1966	16.1	1989	123.8
1967	16.2		
1968	16.8	1990	
1969	17.6	1991	
		1992	
		1993	
		1994	

* Authors' calculation of the implied index of capital goods prices obtained by dividing current price estimates of gross domestic fixed capital formation by the corresponding estimates at 1985 prices. The sources of the primary data are given below.

Sources: *United Kingdom National Accounts* (HMSO annually) and *Economic Trends Annual Supplement* (HMSO).

Appendix B

Names, Addresses and Telephone Numbers

This list provides details of the names, addresses and telephone numbers of the organisations which compile the index numbers referred to in this book.

Association of Cost Engineers, Lea House, 5 Middlewich Road, Sandbach, Cheshire, CW11 9XL. Tel. No. 0270 764798.

Building Cost Information Service, 85/87 Clarence Street, Kingston-upon-Thames, Surrey KT1 1RB. Tel. No. 081 546 7554.

Building Maintenance Information Limited, 85/87 Clarence Street, Kingston-upon-Thames, Surrey KT1 1RB. Tel. No. 081 546 7554.

Building Research Establishment, Bucknall's Lane, Garston, Watford, Herts WD2 7JR. Tel. No. 092 73 76612.

Building Research Station - see Building Research Establishment

Davis, Langdon & Everest, Princes House, 39 Kingsway, London WC2B 6TP. Tel. No. 071 497 9000.

Department of the Environment, Construction Statistics Division, 2 Marsham Street, London SW1P 3EB. Tel. No. 071 212 3000.

Property Services Agency, Apollo House, 36 Wellesley Road, Croydon, CR9 3RR. Tel. No. 081 760 1000.

Scottish Office Building Directorate, New St Andrew's House, Edinburgh EH1 3SZ. Tel. No. 031 244 4173.

Venning Hope Limited, 1 Chartfield House, Castle Street, Taunton, Somerset TA1 4AS. Tel. No. 0823 251417

Glossary

Arithmetic mean

The sum of all the items in a set of values divided by the number of items in the set. It is often referred to simply as 'the average'. It may be defined formally as:

$$\bar{x} = \frac{\Sigma x}{n}$$

where $\bar{x}$ ('x-bar') = arithmetic mean
x = values of the different items
Σx = total of all x
n = number of items

For example, if we have the figures 105, 107, 108, 112 and 118, then:

$$\bar{x} = \frac{\Sigma x}{n} = \frac{(105 + 107 + 108 + 112 + 118)}{5} = 110$$

Average

A summary statistic for describing a set of data. The most common measure is the arithmetic mean (see above). Others are the geometric mean, the median and the mode.

Backdating

Changing a price etc. at a certain date to a relative price at an earlier date (see Chapter 1).

Base year or period

The year or period of time against which all others are compared. Base dates are usually designated as 100 and later dates are designated relative to 100.

Building costs

The costs actually incurred by the builder in the course of his business.

Building cost index

An index based on changes in the costs of the factors of production - labour, materials, plant etc. - employed by builders. It is also called a factor cost index. It should be distinguished from a tender price index and an output price index (see below).

Downturn

A fall, or a reduction in the rate of increase, of prices or the level of activity is an industry or the economy as a whole.

Factor costs

The various costs that make up the total cost, i.e. the factor costs that make up a building cost index are labour costs, plant costs, material costs, rates, rents, overheads and taxes.

Factor cost index

Another term for building cost index. See under that definition.

Fluctuation clauses

Clauses of a contract which allow a tender price to be adjusted after the letting of a contract on account of subsequent changes in materials prices and/or labour costs.

Geometric mean

The nth root of the product of all items in a sample where n is equal to the total number of items in a sample. The geometric mean can be found from the formula:

$$GM = \sqrt[n]{(X_1 X_2 X_3 ... X_n)}$$

where GM = geometric mean
$X_1...X_n$ = each item in the sample.
The geometric mean is often used in index number construction.

Keen

A descriptive term for a tender price or the like which is particularly competitive, e.g. owing to a scarcity of work in a recession.

Laspeyres index

An index which uses quantities obtained from the base year as weights ('base weights'). A Laspeyres price index may be defined as:

$$I_t = \frac{\Sigma P_{it} W_{io}}{\Sigma P_{io} W_{io}}$$

where I_t denotes the index for period t

$\Sigma P_{it} W_{io}$ denotes the sum of the products of the prices in the period under consideration (P_{it}) times base weights (W_{io})

$\Sigma P_{io} W_{io}$ denotes the sum of the products of the prices in the base period (P_{io}) times base-period weights (W_{io})

A Laspeyres index is fairly easy to maintain because the only variable is P_{it} which is the prices in the year under consideration, the prices and quantities in the base year remaining constant.

Market conditions

The economic constraints within which building contractors have to work when compiling their tenders.

Mean

See *arithmetic mean* and *geometric mean*.

Median

A summary statistic for describing a set of data, defined as the middle value of the set when it is arranged in ascending or descending order. Thus the median of the following set of seven price quotations is £8:

£5, £6, £7, £8, £10, £10, £12.

Mode

A summary statistic for describing a set of data, defined as the value that occurs the most frequently. Thus the mode of the set of seven price quotations shown above under 'Median' is £10.

Output price index

An index which measures the cost of construction completed in given time periods. Output price indices are derived from tender price indices and are used as deflators to convert contractors' output of new construction from current prices to constant prices.

Paasche index

An index which uses quantities obtained from the year under consideration as weights ('current weights'). A Paasche index may be defined as:

$$I_t = \frac{\Sigma P_{it} W_{it}}{\Sigma P_{io} W_{it}}$$

where I_t denotes the index for period t.

$\Sigma P_{it} W_{it}$ denotes the sum of the products of the prices in the period under consideration (P_{it}) times current weights (W_{it}).

$\Sigma P_{io} W_{it}$ denotes the sum of the products of the prices in the base period (P_{io}) times current-period weights.

A Paasche index is more difficult to maintain than a Laspeyres index as quantities as well as prices from the year under consideration have to be obtained.

PC sum

A sum inserted in a bill of quantities to cover the cost of work to be carried out by a nominated sub-contractor or for materials to be supplied by a nominated supplier (not to be confused with the PC of daywork).

Preliminaries

Expenses which are incidental to the execution of particular contracts, including site administration, insurances, fencing, site offices, scaffolding, provision of temporary roadways and water supply etc.

Price relatives

The price of an item in one time period relative to another time period expressed with 100 as base. Symbolically it is:

$$\frac{p_t}{p_o} \times 100$$

where p_t = price in the time period under consideration.
p_0 = price in the base time period.

Project index

An index for an individual project obtained by comparing the current tender price with its price revalued using a base schedule of rates. Calculation of a 'project index' is part of the process of compiling tender price indices.

Recession

A decline in the level of economic activity.

Tender price

The price for which a builder offers to erect a building, not including variations, that the client has to pay. It may include fluctuations or it may not depending on the terms of contract. It includes building costs and also takes into account market considerations.

Tender price index

An index based upon data obtained from tender prices (usually from bills of quantities).

Tendering climate

See 'market conditions'.

Updating

Changing a price, etc., at a certain date to a relative price at a later date (see Chapter 1).

Upturn

An increase in the level of activity in an industry or the economy as a whole.

Weighted factor costs

Factor costs which have been quantified as to their relative importance for inclusion within a factor cost index.

Weighting

The adjustment of an input in proportion to its importance.

Yardstick

A standard by which something else is judged.

Index

adjustment for time 14-15
all new construction, output price index, DOE 41
ABI/BCIS house rebuilding cost index **182-85**
APSAB 25, 34
  cost indices **248-59**
Architect and Building News 315
arithmetic mean 359
assessing the level of individual tenders 13-14
Association of British Insurers - see 'ABI'
Association of Cost Engineers 357
  indices of erected plant costs **187-97**
average 359

backdating 13, 359
Banister Fletcher 288, 339
  index of the comparative cost of building **323-27**
base year 359
Baxter indices 157
BCIS 17, 18 199
  ABI/BCIS house rebuilding cost index **182-185**
  building cost index 18
  building cost indices **133-55**
  cost forecast 17
  on-line service 137
  tender price indices **81-101**
BMI
  building maintenance cost indices **198-247**
  property occupancy cost analyses 200
BMI Quarterly Cost Briefing 200

Board of Inland Revenue, index of the cost of construction industrial buildings 288, **314-17**
Board of Trade Journal 335
brick construction cost index, BCIS 133
BRS measured work index 288, **328-331**
Building Cost Information Service 357 - see also 'BCIS'
Builder 300
Building 18, 129, 175
building cost index 360 - see also 'building cost indices'
  BCIS 133
building cost indices 39, **133-259**
  APSAB **248-59**
  BCIS **133-55**
  Banister Fletcher's **323-37**
  Scottish Office **178-81**
  Spon's **156-69**
building costs 360
Building Directorate 178
building maintenance cost indices, BMI **198-247**
Building Maintenance Information Limited 357 - see also 'BMI'
Building Research Establishment 58
Building Research Station 35, 357 - see also 'BRS'
building trades 291-92, 301, 328
Business Monitor MM17 43

capital goods prices 264, 340, 355
Carter, C.F. 334
cash-flow projections 21-24
CCA, CSO index for new building work 310

Index

Central Statistical Office 199, 287 – see also 'CSO'
Chartered Surveyor 315
Chartered Surveyor Building and Quantity Surveying Quarterly 21
Chartered Surveyor Weekly 183
CIOB Code of Estimating Practice 171
civil engineering cost index, Spon's 156
Clark, Colin 305, 306
cleaning cost index, BMI 198
cleaning materials cost index, BMI 198
CNC index, DOE 34, 43, **332-338**
commercial building
  output price index, DOE 41
  tender price index, BCIS 81
comparison of historical indices 339-40
comparison of current indices 261-84
comparisons between published indices 19-20
concrete framed construction cost index, BCIS 133
constructed civil engineering cost index, Spon's 156
constructed landscaping (hard surfacing and planting) cost index, Spon's 156
construction indices
  historical series 287-89, 291-337
  methods of measurement 32-5
  problems of measurement 31-2
  types 7, 10
construction output price indices, DOE **41-55**
construction plant and equipment 40
Cost Engineer 188, 189
cost indices – see 'building cost indices' and 'construction indices'
cost limits 40
'Cost of building chart' 319
Cost of Housebuilding (Girdwood Reports) 306
cost of living index 352
cost planning 16-17, 40
CSO 306

CSO CCA index for new building work 288, 310-13
current cost accounting 310

Davis, Belfield & Everest 287
Davis, Langdon & Everest 357 – see also 'DL&E'
Department of Education and Science 58
Department of Employment 39
Department of Energy 199
Department of Health 21, 58, 59
Department of Social Security 58
Department of the Environment 357 – see also 'DOE'
Department of Trade and Industry 39, 156, 188
Department of Transport 103
derived tender price index 24-29
  for electrical installations 26
  for mechanical services 28
Devons, E. 315
DHSS expenditure forecasting method 21
DL&E 19, 20
  tender price index **128-31**
DOE 33, 39
  construction output price indices **41-55**
  cost of new construction (CNC) index 34, 288, 314, **332-37**, 339
  price index of public sector housebuilding **114-23**
  public sector building tender price indices 25, **57-71**
  road construction tender price indices **102-13**
downturn 360

Economic Trends 43
Economist 300, 306, 315
electrical cost index, APSAB 248
electrical services cost index, Spon's 156
electrical work 136
energy cost index, BMI 198
erected plant costs, Association of Cost Engineers' indices **187-97**

fabric maintenance, cost index, BMI 198
factor cost index 34, 360
factor costs 32, 34, 360
factor costs forecast 18
firm price
  road construction tender price index, DOE 102
  tender price index, BCIS 81
  tender price index, PSA QSSD 72
  tender price index, public sector housebuilding, DOE 114
Fleming, M.C. *viii*, 33, 34, 315
fluctuating price tender price index, BCIS 81
fluctuation clauses 360
fluctuations forecast 17
forecasting 17-18

GDP deflator 264
general maintenance, cost index, BMI 198
geometric mean 360
Girdwood reports 306
Guide to House Rebuilding Costs for Insurance Valuation 182

health service maintenance cost index, BMI 198
heating, ventilating and air conditioning work 136
  cost index, APSAB 248
historical construction indices 287-89, **291-337**
  comparison of 339-40
Home Office 58
House Price Data Book vii
house rebuilding cost index, ABI/BCIS **182-185**
housebuilding - see also 'housing'
  Building, housing cost index **171-77**
  tender price index, public sector, DOE **114**
housing 136 - see also 'housebuilding'
  cost index, *Building* **171-77**
  output price indices **41**

Redfern's index **305**
Scottish Office, tender price index **124-27**
tender price index, BCIS **81**
Housing and Construction Statistics 34, 39, 42-43, 58, 59, 73, 103, 115, 117, 333, 334, 335
Housing Statistics Unit 125, 179
Hudson, K.W. 21

index numbers
  concept 1-7
  historical development 1
  uses of 13-28
index of capital goods prices 355
index of retail prices 353
index of total home costs 354
industrial buildings
  cost of construction, BIR index **314-17**
  output price index, DOE **41**
  tender price index, BCIS **81**

Jones' selling price of building index **291-95**
Jones, G.T. 288, 305, 315

keen (tender) 361

labour cost index, BCIS 133
labour costs 34, 333
landscaping cost index, Spon's 156
Laspeyres index 5, 115, 361
Laxton's Builders' and Contractors' Price Book 292, 293, 296, 300
lift work 135
local authority housebuilding, tender price index, DOE 114
local authority maintenance cost index, BMI 198
logarithmic scales 8-9
London 128, 292, 328
London and Cambridge Economic Service 305, 306
London building industry 289

368 *Index*

maintenance cost index, BMI 198
maintenance materials cost index, BMI 198
Maiwald's indices of building costs and other construction costs **300-304**, 339
Maiwald, K. 288, 301
market conditions 34, 183, 361 - see also 'tendering climate'
materials cost index
  BCIS 133
  maintenance, BMI 198
materials prices 34, 333
measured rates 34-5, 306
measured work index, BRS **328-331**
mechanical and electrical engineering cost index, BCIS 133
mechanical services cost index, Spon's 156
median 361
median index of public sector building tender prices 59
methodology 32-5
methods of measurement 32-5
Ministry of Works 314, 315, 319
MIPS 59
mode 361
Monthly Bulletin of Indices - see 'Price Adjustment Formulae for Construction Contracts'
Monthly Bulletin of Construction Statistics 334

National Association of Lift Manufacturers 40
NEDO 20, 157
Nellis, J.G. *viii*, 33

occupancy cost analyses 199
offices 136
on-line service, BCIS 137
output price index 362
output price indices 39, **41-55**

Paasche index 6, 362
PC sum 362
petrochemical projects 188

petroleum projects 188
PILAH 115, 117
PIPSH **114-23**
plant cost index, BCIS 133
plant, costs of erection - see 'Association of Cost Engineers'
preliminaries 362
price adjustment formula for construction contracts 39, 248, 251
Price Adjustment Formulae for Construction Contracts 20, 134-36, 158, 178
Price Index Numbers for Current Cost Accounting 311
price index numbers for current cost accounting 43, 311
price index of public sector housebuilding, DOE **114-23**
price relatives 362
price trends
  in construction 261-64, 339-40
  in the economy 263-64, 340
prices
  regional differences 73, 82, 117
  locational factors 73, 82, 117
  size of contract 73, 82, 117
  building function factors 73
  houses *vii*
pricing 15-16
principal roads, construction tender price index, DOE 102
Private Enterprise Housing 305
private sector maintenance cost index, BMI 198
private sector tender price index, BCIS 81
producer price indices 199
project index 14, 33, 363
Property Services Agency 357 - see also 'PSA'
PSA 20, 39, 58, 59, 117, 134-36, 157-59, 248-51, 357
  QSSD tender price indices 42, **72-79**
public sector tender price index, BCIS **81**
public works, output price index **48**

QSSD 42, 58, 72

Index

Quantity Surveyors Information Notes 25, 26, 28, 58, 59, 73, 250-51
Quarterly Review of Building Prices, BCIS 82, 83, 137

recession 363
redecorations, cost index, BMI 198
Redfern's indices for housing and for building and works **305-309**
Redfern, P. 288, 305
Report of the Working Party on Building 306
repricing of tenders 33
retail prices index 264, 340, **353**
road construction tender price indices, DOE **102-13**
Robinson, H.W. 306, 315
Royal Institution of Chartered Surveyors 182-83

Sauerbeck index 315
Saville's selling price for building index 296
Saville, J. 288, 296-97
schools 136
Scottish Development Department 58, 125, 179
Scottish Office 103, 125
  building cost index **178-81**
  Building Directorate 124, 357
  housing tender price index **124-27**
services maintenance, cost index, BMI 198
sources, of index numbers 8
South East tender price index, BCIS 81
Spon's Architects' and Builders' Price Book 18, 129, 157, 159, 287
Spon's Civil Engineering and Highway Works Price Book 157, 159
Spon's cost indices **156-69**
Spon's House Price Data Book vii
Spon's Landscape and External Works Price Book 158, 159
Spon's Mechanical and Electrical Services Price Book 157, 159, 189

Spon's Price Book Update 159
Statistical Bulletins 179
Statistical News 116, 117
steel framed construction cost index, BCIS 133
structural steel work 135

Taxation 315
tender price index - see also 'tender price indices'
  advantages 40
tender price indices 39, **57-131**
  BCIS **81-101**
  DL&E **128-31**
  DOE public sector building **57-71**
  housing, Scottish Office **124-27**
  PSA QSSD **72-79**
tendering climate 40 - see also 'market conditions'
total home costs 264, 340, 354
trunk roads, construction tender price index, DOE 102
types of index 39
Tysoe, B.A. *vii*

unit rates 32, 33, 34-5
United Kingdom National Accounts 306, 349
updating 13, 363
upturn 363
uses of indices 13-28

value-weighted tender price index 57
variation of price clauses 20-21
variation of price
  road construction tender price index, DOE 102
  tender price index, Greater London DL&E 128
  tender price index, PSA QSSD 72
  tender price index, public sector housebuilding, DOE 114
Venning 32
Venning Hope 32, 288
Venning Hope Limited 357
Venning, H.J. 305, 315, 319

weighted average 4
weighted factor costs 363
weighting 4-5, 364
Welsh Office 103
Working Party on Building 306

yardstick 364